FISH BEHAVIO[illegible]

in the Aquarium and in the Wild

FISH BEHAVIOR

in the Aquarium and in the Wild

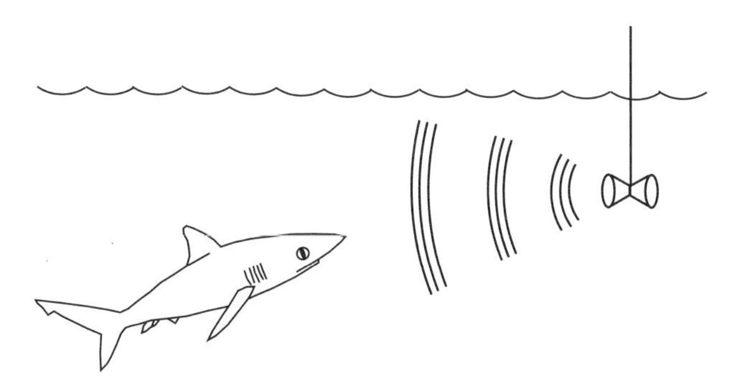

Stéphan Reebs

COMSTOCK PUBLISHING ASSOCIATES, *a division of*
CORNELL UNIVERSITY PRESS
Ithaca & London

First published 2001 by Cornell University Press
First printing, Cornell Paperbacks, 2001

Printed in the United States of America

Library of Congress Cataloging-in-Publication Data

Reebs, Stéphan.
Fish behavior in the aquarium and in the wild / Stéphan Reebs.
p. cm.
Includes bibliographical references (p.).
ISBN 0-8014-3915-9 (cloth : alk. paper)—ISBN 0-8014-8772-2 (pbk. : alk. paper)
1. Fishes—Behavior. 2. Aquariums. I. Title.
QL639.3 .R44 2001
597.15—dc21

2001002593

Cornell University Press strives to use environmentally responsible suppliers and materials to the fullest extent possible in the publishing of its books. Such materials include vegetable-based, low-VOC inks and acid-free papers that are recycled, totally chlorine-free, or partly composed of nonwood fibers. Books that bear the logo of the FSC (Forest Stewardship Council) use paper taken from forests that have been inspected and certified as meeting the highest standards for environmental and social responsibility. For further information, visit our website at www.cornellpress.cornell.edu.

Cloth printing 10 9 8 7 6 5 4 3 2 1
Paperback printing 10 9 8 7 6 5 4 3 2 1

Contents

Preface

Gazing at fishes in an aquarium is like watching a fire or a running stream. It is soothing and slightly hypnotic. This feeling is helped by the quiet disposition of most pet fishes, which are often content to swim slowly, some might say majestically, within their tank. Such tranquility, however, paints only an incomplete picture of the normal behavioral repertoire of fishes. The denizens of the deep do not always go through life at a slow pace. Aquarists who breed aggressive cichlids, or mistakenly put incompatible species together, or drop food in the midst of a hungry group have an inkling that action can get fairly lively in the fish world. Moreover, no serious aquarist can be fooled into the belief that his or her quiet charges would behave beatifically if they were in the wild, having to find food and avoid predators by themselves. In truth there is no reason to expect that fish behavior should be any less diverse, sophisticated, and interesting than that of birds, for example, a class of animal whose behavior has been more thoroughly studied.

With enough patience and discipline, one can observe and document a fairly vast array of behaviors in fishes, either by keeping them in well-developed aquarium installations or by simply following them in the field. There are already a good number of popular and specialized books on the market that describe the natural behavior of fishes—or at least give a glimpse of it, given that there are more than 24,000 recognized species of fishes and that they show great variability in their ecology and behavioral repertoire. In general, we have good descriptions of behavior relating to food searching, predator evasion, fighting, mating, and parental care.

Yet fishes also have special abilities that are not immediately obvious even to patient observers. To investigate some of these abilities, one needs to resort to the experimental approach. For example, to determine whether fishes can hear certain sounds, whether they can discriminate between different times of day, or whether they can distinguish good from bad sexual partners, people must bring the fish into the laboratory and expose them to various artificial conditions. (The really resourceful investigators can even execute their manipulations outdoors and work in the field.) Experimenters must compare the behavior of the fish in control (unmanipulated) and experimental (contrived) conditions, and they must design these experimental conditions in clever ways that will allow them to answer specific questions. Such contrived situations might occur too infrequently in nature simply to hope to chance upon them, even during disciplined and painstaking observations, hence the need for the experimental approach.

Throughout the world, hundreds of professional scientists—I among them—believe that intriguing facets of fish behavior can be demystified through the experimental approach. These scientists call themselves fish ethologists (ethology is the study of animal behavior), behavioral ecologists (ethology with an emphasis on the adaptive function of behavior), comparative psychologists (with an emphasis on mental processes such as learning and memory), or fisheries biologists (with a focus on applied behavior and species of economical importance). All of these scientists are imbued with a passion for what makes fishes tick. All of them spend long hours in the field or more commonly in front of aquaria deep inside the bowels of government laboratories and university halls. There they deviously manipulate the environment of various fish species from the pet trade or from the wild, attempting to test interesting ideas about behavior and all the while trying not to succumb to hypnosis. From these dark basements emerge fascinating discoveries.

These discoveries usually lead to more experiments, more analyses, and more refined interpretations of the results. Eventually, the studies are written up as stylized articles and find their way into the specialized scientific literature. There the knowledge becomes accessible to all colleagues worldwide. Unfortunately, however, the public tends to be unaware of this type of information. This is a pity, given that many of the results are interesting and easy to understand, that the methodology often is not very complicated—many of the experiments could indeed be replicated in a home aquarium—and that a substantial segment of the population—aquarists, fisherfolk, biology students—shares more than a passing interest in fishes.

In this book, I have tried to popularize those facts about fish behavior that are hidden in the scientific literature or are not immediately apparent in the descriptive popular literature. In my mind, this book is aimed at all fish fanciers who, just out of sheer curiosity, would like to know more about the neat things that fishes are capable of doing. It also attempts to bridge the gap between professional ethologists and the public, who, through their taxes and the government agencies that fund such research, make the work possible.

This book is divided into three parts. The first part comprises four chapters under the banner of sensory abilities. Here we explore some of the environmental factors that fishes can detect and the behaviors that are made possible by such powers of perception. These are wonderful abilities, the existence of which we could hardly guess by just looking at a fish in an aquarium or by simply extrapolating from our own sensory capacities. I have covered the topics of olfaction (ours is poorly developed, but perhaps that of fishes is better), hearing (we humans cannot hear much when we play underwater, in swimming pools, for example, but what about fishes?), the lateral line system (what does it do exactly?), and the perception of electric and magnetic fields (not all fishes can do this, but a fair number are specialized in it). I have ignored the visual system because I believe the visual abilities of fishes are not counterintuitive enough or noticeably better than our own—although I must concede that some species can detect ultraviolet and polarized light.

The second part of the book, grouped into four chapters, looks at fishes' cognitive abilities. Here I discuss the special way that fish brains process information, which gives rise to apparently sophisticated behavior. We will not delve into the brain mechanisms themselves—they are poorly known anyway—but rather into the behavioral output. We will see that fishes can learn, can tell what time of the day it is, can recognize many other fish individually, and can somehow assess the danger level associated with a particular predator or rival.

The last part of the book concerns fishes' ability to make adaptive choices. In many species, individuals are capable of carefully choosing their sexual partners. We will review the criteria that form the basis of this decision. In a second chapter we will see that even though all fish shoals may look the same to us, the fishes themselves make fine discriminations when the time comes to decide what kind of shoal they should belong to. And in the final chapter, we will consider interesting examples in which fishes must face contradictory obligations, forcing them to make compromises in their choice of what to do.

When mentioning species throughout this book, I have tried to use common names only. In those instances in which I could not ascertain a common name, I have used the Latin name with a mention of the general group to which the species belongs. I refer readers to the end of the book, where they will find a list of the Latin names that correspond to all of the common names mentioned in the text.

Although the list may appear long at first sight, readers will quickly realize that the same families or groups of fishes tend to appear over and over again in the text, indicating that researchers have favorite species. These are the cichlids, guppies, sticklebacks, minnows, salmon, trout, and some of the most ubiquitous coral reef fishes. Many of these fishes can also be found in the aquarist's home, which shows that an important consideration in an experimenter's choice of subject is whether the animal can thrive in captivity and show normal behavior there.

I have assumed that readers already possess a certain amount of basic knowledge about fishes. Accordingly, I have not always described details of anatomy, general behavioral habits of well-known species, or evolutionary relationships. But for those who would like to delve deeper into the various topics presented, I have included notes that detail the original publications and scientific articles I consulted while researching the book. One should be able to find these sources in a good university library or to get them through the library's interlibrary loan service.

A clarification about two pairs of troublesome words: first, *fishes* and *fish* as the plural form of the singular *fish*. In this book, I have followed the convention that seems to be in favor with most contemporary authors: *fish* refers to one or more individuals of the same species, whereas *fishes* indicates more than one species. In practice, this means that *fishes* will be encountered most of the time throughout the book because I usually discuss this class of animals as a whole, and of course there are more than one species of them.

The other nagging duo is *shoal* and *school*. It used to be that *school* was the word that first came to mind when describing any fish groups, but now scientific convention dictates that *school* be reserved for fish groups that are polarized, with all fish facing in the same direction and moving in synchrony, whereas *shoal* is used as a more general term that applies to any group of social fishes, polarized or not. In writing this book I have almost always used the word *shoal.* I have done this even when designating small groups, in contravention with most dictionary meanings of the word.

I wrote most of this book while on sabbatical from my position as a professor of biology at the Université de Moncton (Canada). I thank the Université de Moncton for granting me this leave and also my host institution, the University of Otago (New Zealand), for making available to me the resources and services of their excellent library. My thanks also to the Department of Zoology at Otago for giving me space and time, and to all of its staff, Robert Poulin in particular, for making me feel very welcome there. For commenting on various chapters, I am grateful to Patrick Colgan, Isabelle Côté, Gene Helfman, Darren MacKinnon, Robert Poulin, Jan Smith, and Catherine Vardy. The reviewing and editing team at Cornell University Press also made very helpful suggestions. Finally, I would like to acknowledge the important role of the Natural Sciences and Engineering Research Council of Canada in supporting my own research on fish behavior.

Part I

Sensory Abilities

1

Olfaction

In 1973, the Nobel Prize Committee surprised the world. The prize in medicine and physiology did not go to a team of cell biologists, geneticists, or physiologists, as had been the case every year until then. Instead, the prize was awarded to three field biologists—Konrad Lorenz from Austria, Niko Tinbergen from the Netherlands, and Karl von Frisch from Germany—for their founding contributions to ethology, the study of animal behavior. All three scientists had worked outdoors with a variety of animal species. They had also studied fishes in the laboratory: cichlids for Lorenz, sticklebacks for Tinbergen, and minnows among others for von Frisch.

Today von Frisch is remembered mostly for his discovery of the waggle dance in bees, but his early work on fishes was no less interesting. During some of his early experiments on minnows, von Frisch needed to devise a technique to mark individuals so that he could recognize them later. In a preliminary trial, he took a European minnow and cut a nerve at the base of its tail, a surgical manipulation that normally leads to a characteristic darkening of the tail. Von Frisch released the freshly operated individual back into its home tank to rejoin its shoal, and then a funny thing happened. All the other minnows within the tank swam away from the marked one and frantically searched for cover. At first von Frisch did not worry too much about it, but then he witnessed the same reaction again some time later when he released another minnow that had gotten stuck under the metal rim of a feeding tube. A hypothesis soon took shape in his mind.

You see, in the eyes of predators, minnows are the perfect prey species. They are abundant, not too big, devoid of spines and toxins, and not overly shy. (It is no accident that minnows are often used as bait in fishing circles.) Nevertheless, minnows do not really appreciate being eaten by other fishes or by birds, and so in the absence of morphological adaptations, they have developed behavioral responses to predation risk. One of them is to live in shoals to better detect and confuse predators. But detection of predators can only be advantageous to a shoal if the first individual to see the threat can somehow communicate its discovery to its shoalmates. This can be done visually of course, but von Frisch surmised that in the case of a minnow that actually got caught by a predator, communication could be chemical. The mangled victim could release a substance that would be carried by water to the rest of the shoal and warn them of the predator's presence. Von Frisch and his students performed a number of experiments to test that hypothesis, and they eventually unveiled the following facts.

(1) The signal is indeed chemical, not visual. Putting an injured minnow into a jar of water for a few minutes, then removing it and pouring only the water at a feeding station into either an aquarium or an outdoor pond cause the minnows there to swim away and not come back for several days.

(2) For the chemical signal to occur, the skin of the injured fish has to be cut. Placing an intact fish in a jar, scaring it somehow, and then using the water from the jar does not induce a

fright reaction in other minnows. Similarly, there is no reaction if the jar holds a fish killed without external injury. Neither do extracts of internal organs elicit a response.

(3) Special cells, called club cells or alarm substance cells, are present within the skin and readily identified under a microscope. They are fragile, rupturing even when a fish merely struggles in moist paper. The substance coming out of them can scare a whole shoal away. In most experiments, 0.002 mg of chopped skin is enough to elicit a fright reaction in a 14-liter aquarium.[1]

(4) This substance was called Schreckstoff (or "alarm substance") by von Frisch. Later one of his students, Wolfgang Pfeiffer, together with the chemists M. Viscontini and M. Argentini from the University of Zurich, identified this substance as hypoxanthine-3(*N*)-oxide.

(5) Minnows detect the alarm substance by smell. Fish cease to react to the substance after their olfactory nerve has been surgically removed; control fish opened up in the same way but without cutting the nerve still react.

The presence of a chemical alarm system is one illustration of the importance of olfaction for fishes.[2] In birds and mammals, warning about predators is done vocally through alarm calls. But in most fishes (suckers, loaches, characins, catfishes, darters, gobies, sculpins, sticklebacks, some cichlids, and salmon, in addition to all minnows), individuals would rather smell danger.[3] The exact nature of the warning substance and the skin cells from which it emanates may vary from species to species; the fright reactions are also very diverse, even within the same population if not within the same individual, but in all cases the response follows the perception of a water-borne chemical by the nose of the fish.

Please Don't Go Away!

The special skin cells that contain the alarm substance are not always present in minnows. When food is scarce, for example, fewer cells are maintained within the skin of minnows, indicating that there is a cost to manufacturing them. It is also common for male minnows to lose their alarm substance cells during the reproductive season. This is an adaptation to their habit of rubbing their body against spawning substrates in preparation for attracting females to a clean nest site. Rubbing causes skin wear, and the subsequent release of alarm substance, if the cells were present, would scare females away, not attract them.

Not all minnows go away when they come in contact with alarm substance, however. Very hungry fish, young fish, and fish that have been exposed too often in the past usually fail to react to alarm substance. Some populations here and there, for unknown reasons, also seem unresponsive.[a]

There is little doubt that alarm substances can help shoal members avoid a predator's operational theater. A simple demonstration of this fact consists of setting minnow traps within a creek, some with a sponge inside that has been dipped in alarm substance (which can be obtained by cutting and washing the skin of a freshly killed minnow) and others with a sponge that has been dipped in distilled water (the control condition). In most cases when this has been tried, the great majority of captured minnows ended up in the control traps rather than in those that gave off alarm substance. One could say that there was something about those treated traps that smelled fishy, and the fish smelled it and kept out.[4]

The effectiveness of an alarm substance can persist beyond the immediate stage of capture by a predator. Indeed, it seems that the alarm substance of a captured prey can survive passage through a predator's digestive tract and emerge still potent in its feces, which can then warn other prey away. In one experiment, three researchers working at the University of Saskatchewan, Grant Brown, Doug Chivers, and Jan Smith, collected the feces of northern pikes that had been fed minnows in their laboratory. (The scientists delicately gathered the feces from the bottom of the pikes' tank, an interesting example of the quaint activities of biologists.) They vacuum-dried these little pieces and suspended them in water. When the researchers introduced this feces-scented water into a minnow tank, the minnows sought cover or dashed about and formed tighter shoals. Other minnows did not react when the experiment was repeated with the feces of pikes that had been fed swordtails, a species that lacks an alarm substance. This control treatment demonstrated that the original fright reaction of the minnows was specifically to an alarm substance and not to anything else that might have been present in the feces.

In a further set of observations, Brown, Chivers, and Smith noted that small pikes housed in a tank 1.83-m long spent most of their time at one end (which had been provided with rocks and artificial plants) but that their feces were found mostly at the other end. It seemed that the pikes were taking great pains to defecate away from their home area. This behavior could be for health reasons of course, but it might also have the additional advantage of locating a prey-warning beacon away from the hunting ground.[5]

The same three researchers also showed that the release of alarm substance could benefit not only shoalmates but also the signaler itself. Their hypothesis was that the release of alarm substance by a prey might attract secondary predators that could interfere with the original hunter and make it drop the prey; the prey might then escape if it were not too badly mangled and if the interfering predator did not get it first. This chain of events would be made possible by the fact that in minnow–pike interactions at least, consumption of a prey is not instantaneous. It takes time for a pike to handle a minnow, half of which is dangling outside the mouth, so that the minnow can finally go in headfirst, apparently a necessary condition for swallowing prey in small pikes. During that time, another pike living nearby could try to steal the prey (in nature, juvenile pikes can live close to one another in weeded areas). In the laboratory, the Canadian researchers, along with colleague Jennifer Mathis, were able to show that secondary pikes are indeed attracted by alarm substance, that they do try to filch the minnow of successful hunters, that pikes harassed in this way sometimes drop their prey, and that such prey are then able to swim at least two pike body lengths away, which presumably would be a safe distance in a heavily vegetated area.[6]

Alarm systems are but one example of the ways olfaction can be used by fish to communicate with one another. Another instance is in attraction of the opposite sex.[7] In a great number of species, females that have ovulated (that is, eggs are present in their ovarian or abdominal cavity) release a fluid from their ovaries, and this ovarian fluid leaks to the outside, where it is perceived by males. The effect on these males is striking. In response, they begin a fairly active search for the source of the odor as well as increase their courtship behavior, nest building, and aggression toward other males. In one of the first reports of this phenomenon, holding water from an ovulated female was poured into the tank of a male frillfin goby, and this male immediately displayed courtship behavior even though his display had to be directed at no one in particular because he was all alone in his tank.[8]

We know that olfaction is involved in this case because the reaction was not observed in anosmic males, that is, males that were smell deprived. (Fishes can be rendered anosmic by severing the olfactory nerve, plugging the nares of their nose with Vaseline, or cauterizing the olfactory sensory receptors. These sensory cells are located in pits in front of the eyes. Destroying them impairs the fish only temporarily, however, because they are known to regenerate in a few weeks or months.) We also know that if the holding water of ovulated females is pumped into one end of a male's tank whereas the other end receives water from unovulated females (see fig. 1.1), the male sets up shop near the "ovulated female" end. Such a preference has been demonstrated in many species but only in normal males. Anosmic males showed no preference for one water input over the other. So, as in better-known cases from the insect world, there exists a sexual pheromone that is broadcast by the ripe females of many fish species, and this substance olfactorily attracts the males to bring the two sexes together and synchronize their reproductive activities.

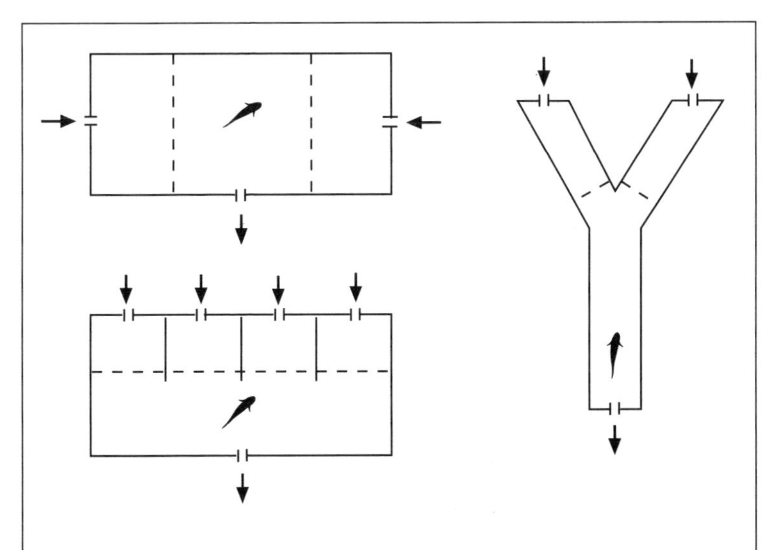

Fig. 1.1. Overhead view of some experimental setups for testing the preference of single fish for different odors presented simultaneously. Arrows show inflows and outflows. Water flow within the apparatus can be checked visually by dripping dyes at the inflow points. Fish express their preference by swimming beyond the dotted lines.

Comprehensive studies on goldfish have revealed the existence of a different pheromone released by females before ovulation. The substance is in fact a female hormone that causes egg maturation, but it can also leak into the environment, where males can perceive it. (Hence, this substance is at once a hormone and a pheromone; a hormone is a chemical that affects specific organs within the body that produces it, whereas a pheromone affects other individuals of the same species outside the body.) The effect of this female pheromone on males is primarily physiological, not behavioral: it causes an increase in milt production. For example, if male goldfish are exposed for only 30 minutes to the holding water of females that are about to ovulate, 6 hours later the sperm volume within their testes will be twice the volume of unexposed individuals. This result is absent in males whose olfactory nerve has been surgically cut, indicating that to be effective, the pheromone has to be smelled.

Olfaction could also play a role in some interesting cases of sexual attraction between species. In North American lakes, the redfin shiner synchronizes its spawning activity with that of the green sunfish. The shiners, in fact, deposit and fertilize their eggs inside the nest and among the clutch of the sunfish. These sunfish are parental: for many days they prevent silting over their nest and aggressively repel predators. The shiner's eggs share the benefit of this parental care. Many species of minnows are known to spawn in the nest of such sunfishes. The question here is: Are the minnows olfactorily attracted to the breeding sunfish?

John Hunter and Arthur Hasler from the University of Wisconsin have shown that redfin shiners are drawn to the nests of green sunfish and are induced to spawn in them by the presence of sunfish milt and ovarian fluid rather than by the sunfish themselves. By siphoning from a bucket, these researchers released either milt or ovarian fluid into the corner of a pond devoid of sunfish; soon thereafter, shiners gathered around the nozzle of the siphons and initiated normal spawning behavior, laying their eggs right there over the muck, in the complete absence of any real sunfish. The researchers also released a lone male sunfish into that pond and let him build a nest, but of course being alone, he did not secrete any milt, and the shiners did not conduct any spawning activity near his nest. So it is not the male sunfish or his nest per se that attracts the shiners but instead his spawning secretions. As in intraspecific sex attraction, detection is more than likely through olfaction.[9]

Olfactory communication does not necessarily require the production of specialized chemicals such as alarm substances or pheromones. The ordinary smell from skin mucus can help individuals of the same species find one another and stay together, which could be useful at night or in turbid water when vision is limited. We can test this experimentally by placing a fish in the middle of a tank with compartments at either ends delimited by pierced partitions, followed by placing conspecifics in one of those compartments but installing no fish—or fish from another species—at the other end. We will note that the middle fish prefers to hang out near the compartment with conspecifics. We can eliminate the possibility that conspecifics are detected visually by using various techniques: working in complete darkness, or using opaque partitions pierced by tiny holes, or fitting foil eyecaps over the eyes of the choosing fish. This would leave only olfaction as the main mechanism underlying side preference. The final proof would be to test anosmic fish and show that they do not exhibit any side preference, whereas normal individuals do.

Experiments of this kind have been conducted with various salmonids, minnows, herring, carp, catfish, zebrafish, gouramis, and rudd, among others. Each time, good evidence for olfactory detection of conspecifics was obtained. In a different kind of experiment, carps were first exposed to a simple wooden model of a fish, and they blissfully ignored it; when the model was coated up with carp skin mucus, however, the carp followed it around. Therefore, it seems that mucus can impart a specific odor to the body it covers, and this odor is enough to attract conspecifics.[10]

An intriguing observation by John Todd suggests that more detailed information could be conveyed by the smell of mucus. Todd was pursuing graduate studies at the University of Michigan in Ann Arbor, under the direction of John Bardach and in collaboration with other scientists, all of whom were working with bullhead catfish. Catfish are normally territorial, but when they are forced to cohabitate at high densities, they are able to live in harmony as a group. Todd and his co-workers had a pair of catfish that shared the same aquarium. The catfish had established territories in different parts of the aquarium, and they often fought with each other. The scientists then pumped into their aquarium the holding water of a large group of peaceful catfish living in a nearby tank. After a few days of this treatment, the protagonists stopped fighting. The water input was then interrupted, and the fighting resumed after a day or so. The water was pumped again, and the cease-fire came back into effect the next day. Todd and his colleagues postulated the existence of a "love-in" pheromone emanating from the communally living bullheads, one that could be olfactorily perceived and that would have pacifying effects not only on the members of a group but even on territorial individuals. (This work was performed in the late sixties; if one wishes, the more technical term *appeasement factor* can be used in place of the hippie-sounding *love-in pheromone*.)[11]

I once did a little experiment to show that even the scent of eggs could serve an important function, at least for the parents. I kept pairs of convict cichlids in the laboratory. Like many New World cichlids, convicts lay eggs on a solid substrate (mine used the inside of a clay flowerpot), and for several days the female takes care of those eggs by "fanning" them, using fin movements to circulate well-oxygenated water over the eggs. With a pair of infrared goggles I had bought at an army surplus store, I noticed that female convicts keep on fanning throughout the night. But how could they find their eggs in complete darkness? I knew from previous studies that they could not see the infrared light from the special flashlight I was using, and I had even taken the trouble to disable the pilot light in all water heaters.

In the middle of the night, working with my infrared goggles, I substituted the flowerpot in which a female had laid her eggs for various experimental pots. If the new flowerpot was covered with little drops of wax the same size and in the same arrangement as the original eggs, the female did not fan them in the dark. But if I placed the female's eggs in an empty tea bag (which prevented direct contact but allowed odor diffusion) and taped that tea bag to the new pot, the female fanned the tea bag just as much as she would have her normal clutch (see fig. 1.2). I concluded that female convicts use chemoreception, and not tactile stimuli, to detect the presence of their eggs in the dark.[12]

Using Near-Infrared Equipment to View Fishes in the Dark

For many years now, I have had fun spying on my captive fishes at night. This can be done under very dim illumination, but for a better idea of what fishes can do in nature, where complete darkness usually prevails, I find it preferable to rely on near-infrared technology. I use a pair of infrared goggles I bought for a cheap price at an army surplus store. A strap allows me to wear the contraption on my head, leaving both hands free for taking notes.

The term *near-infrared* refers to the wavelength that is just beyond visible red. Near-infrared technology therefore requires a source of infrared light to illuminate the subject. A flashlight fitted with an infrared filter does the job nicely. *Far-infrared* refers to the wavelengths that correspond to heat emission. Here a light source would not be necessary because the subject itself would be generating the heat. Far-infrared technology can be used for night viewing of birds and mammals but obviously not fishes, which are ectotherms (that is, they are cold-blooded) and do not give off much heat. Unfortunately, even near-infrared technology is not without drawbacks for observing fishes. The main problem is that water absorbs infrared light very readily. This means that subjects can only be illuminated through a relatively thin layer of water. Using more powerful lights does not necessarily

Fig. 1.2. Inside her nest—a bottomless flowerpot lying on its side—and in the complete darkness of night, a female convict cichlid fans a clean tea bag that contains her eggs, put there by an experimenter. She takes care of that tea bag just as she would a normal brood of eggs. This experiment shows that she can find her eggs by using chemical cues and that tactile cues are not important.

solve the problem, because visible red light may then bleed through the filter, or the heat produced by the light may burn off the filter if it is made of gelatin. I use flashlights of medium power and resign myself to viewing only the front 10 cm of my aquarium.

I have used this system mainly to observe cichlids. I am lucky in that physiologists have already done the hard work of showing that the eyes of cichlids cannot pick up infrared light (they are therefore truly in complete darkness when I look at them). Cichlids are very parental, taking assiduous care of eggs (embryos), wrigglers (larvae), and free-swimming fry (very young juveniles). It was neat to discover that they keep up their parental duties at night.[b]

To most nonparental fishes near the nest of my cichlids, egg odor would have had a very different meaning. It would have signaled food. Olfaction is indeed very important for food finding in fishes.[13] Anglers, for example, are well aware of the power of chemical attractants and odorous bait as a way of drawing fish to lures, traps, or specific areas. There have been a great number of studies on this topic. The fishes themselves show great variability in their response to food odors and in the sensory modalities they use for detection. Some species rely on olfaction, others use their sense of taste, and some exploit both. Even closely related species may differ: for example, the brown bullhead uses mainly olfaction to find food, whereas the yellow bullhead uses mainly taste. It is often necessary to lump olfaction and taste together and apply the more general term *chemoreception* to any discussion of the effect of food odors on fishes.

Most studies have employed before-and-after protocols: they compared the behavior of the same fish before and after the introduction of water laced with food odor. In general, fishes simply start to swim more actively after "food" water is introduced into their aquarium. In nature, detection of the odor incites many species to swim upstream, a response called positive rheotaxis. This is a reasonable strategy used to get closer to the source of an odor. Sharks, for example, are often reported to approach bait from a downstream direction. In the absence of currents, some species adopt the strategy of swimming in a straight line as long as they perceive an increase in odor intensity (which can only mean that they are getting closer to the food), but they turn left or right at random when intensity fades out. The resulting trajectory is a zigzag that ultimately brings the fish closer to the food. This orienting mechanism is called klinotaxis.

Another search pattern, called tropotaxis, is only possible in fishes whose left and right receptors are widely separated. These fishes can compare the stimulus intensity on the left and right sides of their body and turn in the direction of the strongest stimulation. Catfishes use the taste receptors on their barbels and flanks in this way. When researchers cut the barbels and denervated the flank receptors on one side of a catfish and then placed this catfish in the chemical field of a food odor, they observed that the catfish kept turning toward its intact side, where the stimulus perception was obviously stronger. This turning behavior was so pronounced that the fish ended up swimming incessantly in circles.[14]

Food extracts can also trigger foraging (food-searching) behavior. When chemically aroused by the scent of food, bottom-dwelling species start digging into the substrate or moving their snout along the bottom. Sight feeders start striking at various small objects whether they look like food or not. Juvenile salmon in rivers snap at drifting tidbits that they would normally ignore. Each species adopts its own typical foraging behavior, and the rigidity of this behavior can sometimes lead to comical situations in the laboratory, as in the case of a bottom-dwelling sole that doggedly searched along the bottom while being stimulated by the smell of a piece of food suspended just above its head. The sole just could not figure out why it never found that delicious-smelling snack that was obviously so near.

Chemoreception: The Difference between Olfaction, Taste, and Oxygen Sensors

Fishes can detect chemical substances with their sense of olfaction or taste. In the case of olfaction, the sensory cells are located at the bottom of two pits (one on the left and one on the right) in front of the eyes. Each pit is connected to the outside by a single nare or by a pair of them. Water is moved through the nares and over the sensors by natural currents, the forward motion of the fish, or rhythmic changes in pit shape caused by the respiratory movements of the nearby mouth. A specialized olfactory nerve links the sensors to the brain.

The olfactory system is anatomically distinct from the sense of taste, which is based on taste buds located in the mouth and often on the body itself. The body surface of catfishes is known to hold as many as 200,000 taste buds, many of them on the barbels; sensitivity tests have shown that catfishes are blessed with one of the most exquisite senses of taste in the whole vertebrate world.[c] The taste buds are connected to special areas of the brain via major nerves that are not solely devoted to gustatory information. Taste buds may or may not respond to the same substances as the nose.

In addition, special sensory cells that belong to neither the olfactory nor the taste system can be found in the mouth of fishes and are sensitive to dissolved oxygen levels. When dissolved oxygen levels dip below a certain threshold, fish often react by increasing their rate of gill ventilation (easily witnessed as an increase in the rate of gill cover opening), by breathing very near the surface (where water may contain more dissolved oxygen because of its proximity to air), or by outright air breathing (some fishes can gulp air and then use their mouth, esophagus, swimbladder, or even stomach like we do lungs, as a site of gas exchange between air and blood vessels). Parental species that provide oxygenated water to their eggs with rhythmic movements of their fins can be experimentally coerced into fanning more by connecting a tube to their nest and pumping oxygen-poor water through it. Similarly, many parental cichlids that normally would keep their young in pits dug in the sand are prompted by low oxygen levels to suck the young into their mouth and spit them into vegetation where the young can tap into the oxygen released by plant photosynthesis. The young affix themselves to the plants thanks to glue-producing glands on their heads, and therefore the behavior is known as wriggler hanging.[d]

As humans, we can relate to the power of odors in eliciting appetite. But the next example of an olfactory function should not have the same familiar ring to it, at least not for those of us who keep a clean house: fishes follow their nose when the time comes to find home.[15] First, on a broad scale: young fish may learn the smell of the general habitat type in which they were born and use that smell as a cue for habitat choice when the time comes to settle down as an adult.

For example, anemonefishes are a popular marine species in aquaria. They live in close association with sea anemones. In nature the fish lay their eggs on a rock close to their host anemone, and after hatching the young spend a little time next to the anemone. Soon thereafter the young leave and spend several weeks growing in open waters. At the end of this period, they must find and settle inside their own anemone. In general, anemone choice is species-specific, that is, each species of anemonefish is always found in its own typical species of anemone. How does each fish recognize its species-specific host anemone among all of the different types that exist? Does it have an innate representation of what the right anemone looks or smells like? Or does it learn this representation soon after hatching next to the parental anemone, just before becoming pelagic?

It turns out that a learning process is involved and that anemone odor is important. The importance of learning is illustrated by the fact that anemonefish that hatch in an aquarium devoid of anemone have trouble finding anemones later in life, and if they are forcibly left in the presence of one, it takes them a fairly long time—almost 2 days—before they settle in it. In contrast, fish that were born next to an anemone quickly find one of the same species, even if it is hidden behind a screen, and they settle in it within 5 minutes. They do not react to other species of anemones.

Interestingly, if these normal fish are put inside an aquarium that receives an inflow of "anemone" water at one end and that offers a view of an anemone at the other end (just a view, no olfactory contact is possible because the anemone lays in an adjacent aquarium), the fish spend almost all of their time next to the inflow, ignoring the anemone in plain sight at the other end (see fig. 1.3). Another piece of evidence for the role of olfaction in this system is that if anemonefishes are

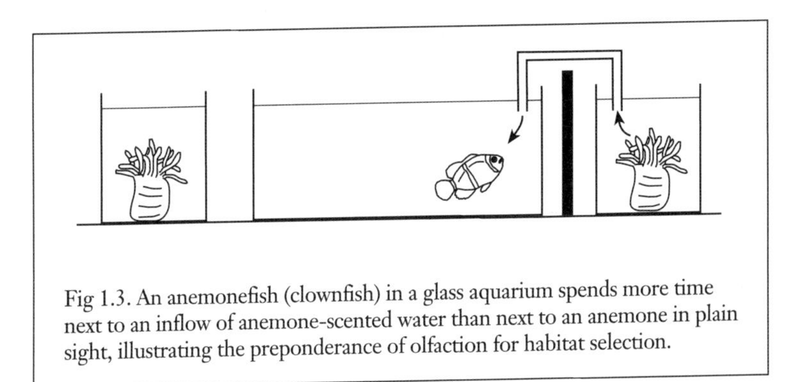

Fig 1.3. An anemonefish (clownfish) in a glass aquarium spends more time next to an inflow of anemone-scented water than next to an anemone in plain sight, illustrating the preponderance of olfaction for habitat selection.

placed in a big aquarium with a current, they have a much easier time finding an anemone that is located upstream rather than downstream or to the side. Such laboratory experiments indicate that anemonefishes recognize their species-specific host anemone by smell and that they learn this odor after hatching next to the anemone of their parents.[16]

Like anemonefishes, damselfishes go through a drifting larval stage before settling down, in their case on a patch of coral. A number of studies have shown that the presence of other residents on coral heads can increase the probability that larvae will settle there. In one field study conducted by Hugh Sweatman on the Great Barrier Reef of Australia, water was pumped at night from coral heads with and without resident humbug damselfish into similar but unoccupied coral patches. The larval fish that settled during the night on these scented patches were collected at dawn by spreading the anesthetic quinaldine over the coral. At the sites supplied with humbug-occupied coral water, many larvae of the humbug damsels ended up settling. At the sites supplied with unoccupied coral water, fewer larvae settled. Sweatman concluded that the odor of adult conspecifics acts as an attractant in the settlement behavior of larval humbugs. (This assumes that the odor of the coral itself did not differ between occupied and unoccupied sites.) He pointed out that chemical information might be important for all species, such as humbug damselfish, that settle at night and cannot visually inspect their prospective real estate.[17]

On a finer scale, fish could remember the smell of their specific home within a given habitat and use it to find their way around if they happen to get displaced by a storm, a predator attack, or inquisitive scientists. For example, Morten Halvorsen and Ole Stabell from the University of Tromso in Norway displaced brown trout 200 m upstream or downstream from their home site. Beforehand, they had anesthetized the fish and cauterized the olfactory sensors of some of them while cauterizing two sites near the nasal openings of the others (this latter procedure did not make the fish anosmic but provided a control for the possible shock of operation and handling). Then, for the next 9 weeks, the researchers patiently waited near the home site to see who would come back.

More than four times as many control fish came back to their home as compared with the smell-impaired individuals. Interestingly, the controls that had been displaced upstream, where they could not smell home, came back just as successfully as those that had been moved downstream. We can therefore imagine that within a stream habitat, displaced fish might follow a rule such as "If you can smell home, swim against the current until you recognize where you are, but if what you smell is not like home, then swim with the current until you do smell home." Halvorsen and Stabell also went back to the upstream and downstream sites of release and found that most of the anosmic trout had remained there, so maybe the general rule for those fish was "If you can't smell anything, stay put."[18]

It is a short jump from trout to salmon, which provide the classical example for the importance of olfaction in homing behavior. When they are young, salmon learn the smell of the stream in which they live. Later, they leave the stream to go and eke out a living at sea. Much later still, they come back to their natal stream to spawn. These adult salmon find their natal stream by following the trail of the odor learned several years earlier but not forgotten. The main player in the research that elucidated the mystery of this homing mechanism was Arthur Hasler from the University of Wisconsin (the same Hasler who had studied redfin shiners spawning in sunfish nests). The time frame was the 1960s and 70s. Hasler's first experiment followed the now familiar protocol of catching adult salmon, making some of them anosmic (by plugging their nostrils) while leaving others untouched, and measuring their respective success at returning to the stream in which they had been born and originally tagged. As one can guess, fish with unplugged noses made it home successfully, whereas the anosmic ones were recaptured more or less evenly among all of the streams of the basin.

The point was driven home in a subsequent series of elegant experiments. Hasler and his co-workers reared young coho salmon in a hatchery and exposed them to one of two different chemicals, morpholine and phenethyl alcohol (PEA). These artificial chemicals do not normally carry biological meaning but they are odoriferous. The fish were then marked according to the chemical they had been exposed to and released into Lake Michigan. During the spawning migration 1.5 years later, the researchers dripped morpholine into one river and PEA into another 9 km away. Convincingly, 95% of the fish that were recaptured and that had been exposed to morpholine were recovered in the morpholine-scented river, and 92% of the recaptured PEA fish were recovered in the PEA-scented stream. One cannot ask for a better experimental demonstration of the importance of odors for homing salmon.[19]

We still do not know the nature of the chemicals that provide the odor learned by fishes in the wild. Geosmin, a chemical produced by tiny mushrooms and present in inland waters, is one possibility. Glass eels have been shown to detect it and to prefer water laced with it at a time when they want to migrate into rivers.[20] Another intriguing possibility is that the smell of other fish might contribute to the olfactory "bouquet" of the home stream. Experiments have revealed that salmon can distinguish between the smell of conspecifics from their native population versus that of unfamiliar populations. One of these experiments showed that Arctic char that were reared in hatcheries with some of their brothers and sisters, whose smell could therefore be memorized, preferred, once released in the wild, to ascend the river in which other relatives were present.[21] So maybe adult salmon migrating upriver could in fact be seeking the smell of the juveniles presently living in their natal stream, if we are willing to assume that these adults and juveniles all share the same family scent.

We have seen that odors—be they detected by olfaction or by taste—are very important to fishes. They are so important in fact that some smelly compounds have already found commercial application. For example, in aquaculture, artificial feed can be made more palatable by the addition of amino acids that are known to stimulate feeding when fishes smell them. Many studies have already demonstrated that farm fishes feed more and grow better on such "flavored" diets.[22] Successful reproduction in farm fishes could likewise be bolstered through the use of specific pheromones.

Unfortunately, enthusiasm for the great potential of chemical stimulants in the fish industry must also be matched by concern for the effects of industrial pollutants on the olfactory world of wild fishes. Heavy metals have been shown to be particularly nasty, constantly destroying sensory cells in the nose of fishes and disrupting most of the behaviors that are known to be based on olfaction.[23] Fishes actively avoid contaminated water in the laboratory, but in nature they may not always have a choice. Fishes live in a world in which all dissolved substances come into intimate contact with their bodies, with great impact on their lives. Everything soluble that we dump into the water has the potential to be perceived by fishes and to affect them, sometimes in unfavorable ways.

2

Hearing

Until the late 1930s, many people believed that fishes could not hear. After all, anyone could see that there was no hole in the side of a fish's head to provide a connection to the internal ears. Dissection had also indicated that these inner ears did not contain a cochlea, the anatomical structure that allows humans to hear. Moreover, anyone who had ever been underwater, if only in a backyard swimming pool, knew that only muffled sounds could be heard there, giving the impression that water distorted sound.

There were, of course, valid counterarguments. It was known that the soft tissues of fishes had almost the same density as water, which meant that underwater sounds could travel through the body of a fish and therefore there was no need for a direct connection between the inner ear and the outside. Physicists could also explain that water was incompressible, which meant that sound could travel very fast and very far in water. So, although the properties of sound propagation were not the same in water as in air, the aquatic environment could still provide its occupants with a rich source of acoustical information. Yet these were only arguments; up to that point, fish hearing had not been convincingly demonstrated.

However, in the 1920s and 30s, Karl von Frisch published an overview of several experiments he had conducted in his laboratory on the hearing capacity of fishes. (He published in German in 1923 and in English in 1938.)[1] For his demonstration that fishes could hear, von Frisch took advantage of the presence of a blinded brown bullhead in his laboratory. That catfish spent most of its days inside a clay pipe. When von Frisch dropped food in the water, the fish smelled it and came out of the shelter to grab it. After a while, von Frisch tried to train the fish by whistling a few seconds before dropping the food. After a few such pairings of sound and food delivery, the catfish started to come out of the shelter as soon as von Frisch whistled, before the food fell into the water. Sound had become linked to food in the animal's mind. Obviously, for this to be possible the fish had to be able to hear the sound.

Even nowadays this kind of training (called Pavlovian conditioning, after the Russian physiologist Ivan Pavlov, who played the same trick on dogs) remains the method of choice to determine the hearing abilities of fishes. Signals can be sounds of different frequency (pitch) or amplitude (loudness). They can also be long sounds that suddenly change in frequency, amplitude, direction, or distance, to see if the fish can discriminate between various levels of these variables. The conditioning procedure in this case would be to offer food only after a sound that includes a shift. If the fish can learn to respond only after such changing sounds and not after stable ones of similar duration, then we can only conclude that they can tell the difference between the two levels that characterize the change.

The main conclusion to emerge from experiments of this kind is that fishes hear well enough, but their perception seems to be biased in favor of low-pitch sounds.[2] Most fishes can hear only in the range of 30–3000 Hz (compare this with an approximate range of 30–16,000 Hz for humans). Each species has a frequency that it perceives particularly well, reacting to it even when the signal is relatively faint. As one gets away from this frequency, however, it takes tones that are louder and louder for the fish to be able to detect them. The reason for this differential sensitivity is that hearing in fishes depends on many anatomical structures that vary in density, shape, and resonance. The mechanical constraints associated with each one of these structures determine which frequency is heard easily and which one is not.

Each inner ear holds three sacs, and within each sac is a small "stone" of inorganic salts—called an otolith (or otoconia in sharks and rays). Each otolith sits on a gelatinous membrane. This jelly encases hairlike projections from sensitive cells that are packed within the sac wall (see fig. 2.1). The otolith does not have the same density as the surrounding soft tissues, and so it does not vibrate in the same way when sound travels through the body. The motion of the otolith relative to the wall of the sac in which it lies bends the hairs that

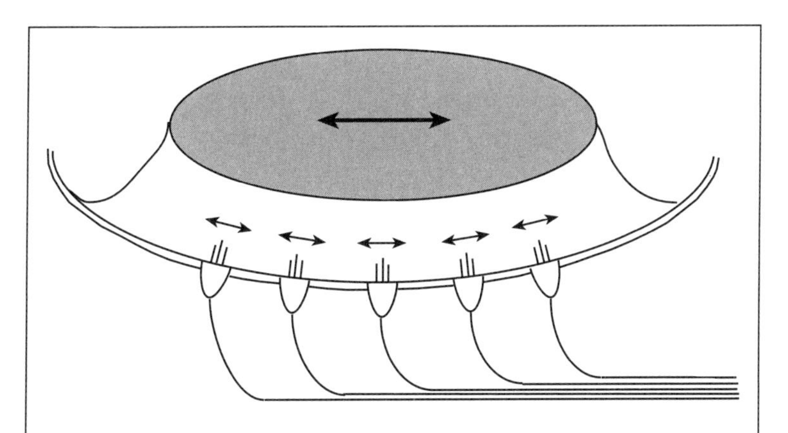

Fig. 2.1. The inner ears of fishes contain a number of otolith organs, one of which is shown here. The otolith (in gray) is a stone of inorganic salts sitting upon a gelatinous membrane. Embedded within the membrane are hairs from sensory cells, a few of which are shown here. Sound travels through the body of the fish and makes the otolith vibrate, which moves the hairs, which in turn causes the cells to fire a signal to the fish's brain.

connect the two. Bending the hairs makes the sensitive cells fire a signal to the brain.

Using Otoliths to Establish the Age of Fishes

The otoliths, along with the rest of the fish's body, go through a cycle of fast growth in summer and slow growth in winter, resulting in a concentric pattern of light and dark bands, much like the growth rings on the cross-section of a tree trunk. Otoliths, therefore, have become an important tool for measuring the age of fishes. The width of the bands can also reflect switches between habitats that are more or less favorable for growth. And in juvenile fishes, even the difference in growth between day and night can be distinguished, so age can be determined in number of days, and daily growth conditions (for example, food availability and impact of a temporary disease) can be guessed at.

Surprisingly, this hearing mechanism is assisted by an organ that does not belong to the auditory system per se—the swimbladder.[3] This gas-filled bag is devoted primarily to floatation and depth adjustment, but it can also act as a transducer, transforming pressure waves from the water into pulsating movements, a motion that can be transmitted to the ears. The importance of this effect stems from the physics of sound travel in water. Sound carries through water in two forms: as a particle displacement wave and as a pressure wave. As a somewhat imperfect analogy, we may think of a wave traveling at the surface of water: the water molecules at the surface move up and down—that is, like particle displacement—and the weight of the water over a given point also goes up and down—that is, somewhat like pressure. Particle displacement waves are quickly attenuated and cannot carry far. In contrast, pressure waves can travel very far.

The otoliths work on the principle that they vibrate out of phase with the rest of the body. Vibration means movement, and therefore the otoliths are built to detect particle displacement, not pressure. The swimbladder, for its part, is filled with compressible gas, which means that pressure waves create cyclic changes in swimbladder shape. Therein lies the transducing action of the swimbladder, from pressure to movement, and this movement can be transferred to the otoliths. Fishes without a swimbladder cannot take advantage of the pressure aspect of sound, and therefore they cannot perceive sound from a great distance because only pressure waves carry far.

The vibration that is transmitted from the swimbladder to the otoliths can travel through the intervening body tissues or via more direct connections. For example, in the superorder Ostariophysi—minnows, catfishes, electric eels—there is a chain of three small bones called Weberian ossicles, whose sole function is to link the swimbladder with the inner ears. In other fishes such as cod, squirrelfishes, and porgies, the swimbladder touches the skull directly. In herring and related species, the swimbladder even makes inroads into the skull and comes into contact with a fluid-filled cavity next to the ears. In cod, goldfish, and catfish, surgical deflation of the swimbladder greatly impairs hearing ability.[4] And in the common dab—a flatfish that lacks a swimbladder—placing a substitute swimbladder (a small inflated balloon) close to its head results in a notable improvement in hearing sensitivity.[5] (There's a good cartoon, Far Side style, somewhere in here: a shoal of old fish swimming about with balloons stuck to their heads as hearing aids.)

Pavlovian conditioning experiments with suddenly changing sounds have revealed that the best listeners, usually species with Weberian ossicles, can tell apart frequencies that differ by as little as 3% (that is, a quarter tone), amplitude variations of 3 dB, and directional angles of 20°. In general, humans can do much better in all of these categories. However, we should not become too smug, because some fishes are better than us at recognizing individual pulses that are separated by very short pauses: they can perceive staccatos that we would only hear as whistles. They can also distinguish sounds that come from directly ahead versus directly behind, or from straight above versus straight below, better than we can.

The mechanisms underlying acoustical discrimination are well known in humans but poorly understood in fishes. In people, frequency discrimination depends on the cochlea, a tapered and rolled-up tube with sections that each resonate at different frequencies of sound. The brain can tell which frequency is being heard by identifying which section of the cochlea is vibrating. As mentioned earlier, however, cochleae do not exist in fish ears. For directional information, the human brain relies on differences between the sound perceived in the right and left ears—in the arrival time of a sound and in its amplitude. The more to the right a sound comes from, the earlier it arrives at the right ear as compared with its arrival at the left one; the sound is also louder on the right side, with the head providing a sound shadow for the ear on the other side. The same mechanisms theoretically could apply to fishes, but this seems unlikely, first because sound arrives almost simultaneously at both ears no matter where it comes from (sound travels fast in water, and fish ears are very close to each other), and second because fish heads are of a similar density to water and do not provide much of an acoustical shadow. A better hypothesis is that the frequency and direction of a sound determine the frequency and direction of hair-bending inside fish ears, and the brain can measure these parameters.

What kind of natural sounds are important to fishes? Historically, research on this topic got a boost from a surprising direction—the advent of World War II. The war fostered important developments in underwater listening technology to better detect enemy submarines, and thereafter, biologists were able to "borrow" the technology. After years of making do with barely adequate equipment, ethologists could at last use hi-fi hydrophones to record natural underwater sounds, and they could install underwater speakers to play back various sounds to fishes to measure their reactions. Many experimental manipulations thus became feasible.

Even so, scientists were not completely out of the woods. One problem was, still is, that small aquaria do not lend themselves well to acoustical studies because their walls reverberate sound and corrupt the purity of the signal. Large acoustically insulated tanks have to be built. (Incidentally, marine species seem better represented in this field of research than in others. Perhaps it is because marine fishes are usually kept in larger tanks or live in more open areas in nature.) Another problem is that the sea is a noisy place—crashing waves, currents, boat traffic, singing whales, even the swimming movements of fishes—and it is hard sometimes to record a given sound clearly or to evaluate how well a fish can hear playbacks over the din of natural surroundings. But with some care, these problems can be, and have been, overcome.

With a good hydrophone, one notices very quickly that fishes can do more than just listen for sounds; they can produce them as well. To do this they knock or rasp some of their teeth, spines, and bones together. They can also use their swimbladder like a drum, making it vibrate through the rapid contractions of special muscles. (Swimbladders, therefore, are useful not only for sound detection but also for sound production.) In the oyster toadfish, the swimbladder muscles are responsible for the production of a "boatwhistle call." They contract at a rate of 200 Hz—that is, 200 individually defined contractions per second—making them the fastest contracting muscles of the vertebrate world. Second place goes to the shaker muscles in the tail of rattlesnakes, which contract at only half that rate.

The sounds produced by fishes are often described as grunts, croaks, hums, moans, thumps, pops, buzzes, clicks, howls, knocks, or snores. As most of these labels suggest, fishes produce sounds of low frequency. Remembering that the hearing ability of fishes is also biased in favor of low frequencies, we quickly realize that sound production might be used by fishes to communicate with one another. In this, some fishes show a remarkable resemblance to birds, a kind of animal that is much more readily associated with the notion of acoustical communication than are fishes.

In birds, males sing to advertise territory ownership, attract females, and warn other males away. In many of the fishes that are territorial during the reproductive season, vocalizations seem to fulfill the same functions.[6] In species such as the mormyrid fish *Pollimyrus isidori* and the intertidal plainfin midshipman, specific sounds are produced only by territorial males, the same as in birds. In other species such as the longspine squirrelfish, territorial calling reaches a peak during the twilight hours, another characteristic shared by birds. Territorial vocalizers can be found among nocturnal species (the examples above) as well as diurnal ones (sunfishes, cichlids, and damselfishes). Acoustical advertisement seems common in fishes that spend a lot of time hidden in crevices and holes, such as the frillfin goby and the oyster toadfish.

A Loud Neighbor

The plainfin midshipman, *Porichthys notatus,* is called the California singing fish by fishermen. During the reproductive season, males "hum" to attract females. A single individual can hum for as long as an hour without pause, easily the longest uninterrupted fish sound ever reported. These fish are nocturnal like frogs, and as in frogs, their combined output can create quite a racket. So much so that in San Francisco Bay, the California singing fish is sometimes the focus of complaints by houseboat dwellers who cannot sleep because they can hear the fish's chorus through the bottom of their boats.

As in birds, vocalizations can be a part of courtship. In numerous species, males can be overheard vocalizing while they court females. They also increase their rate of "singing" when they see females approaching. Females, for their part, are attracted to these serenades. In several experiments, females were observed investigating speakers that played back male calls. They ignored playbacks of other sounds. Interestingly, nearby males often reacted to the playback of a rival's courtship calls by increasing their own rate of vocalizations, or their own rate of courtship displays, as if to better compete for the females' attention.[7]

As in birds and frogs, females can use various characteristics of individual calls to assess the quality of males. One example comes from the work of Arthur Myrberg, an expert on acoustical communication in fishes. Working with the bicolor damselfish off the Florida coast, Myrberg and his co-workers measured the reaction of females to the playbacks of individual males. The researchers set up a line of big conch shells near a colony of damselfish and hid speakers behind them. They selected some of those speakers more or less at random and played back the calls of either a small male or a big one. They observed that nearby females preferred to investigate the conch shell (speaker) that played the big male's calls. Females usually prefer to mate with big males, and it seems that they could tell the difference just based on acoustical cues.[8]

Interestingly, cruising trumpetfish, a predator of damselfish, also inspected the speakers that played male calls, while passing over speakers that played only white noise. Therefore it seems that male damselfish can use vocalizations to advertise their good size and availability, but that they may also pay a price by revealing their position to predators. Similar situations have been documented in tropical frogs in which females prefer males with deeper voices and frog-eating bats attack calling males.

The habit by territorial males of responding to the calls of other males can be put to good use in tests of recognition. Here again the work of Myrberg is enlightening. Working in the Bahamas, he and co-worker Juanita Spires tested the ability of the bicolor damselfish to distinguish between their species-specific courtship calls and the very similar vocalizations of two closely related species, the beau-gregory and the spot damselfish. The researchers observed that the bicolors reacted with more vocalizations and more courtship displays to playbacks of their own conspecific calls than to vocalizations from the other species. In subsequent research, Ehud Spanier, a Ph.D. student on Myrberg's team, showed that the other two damselfish species also respond more to their own specific calls. Minute differences in pulse intervals and pulse number within each call seem to be the key features involved in specific recognition.[9]

Myrberg and Robert Riggio carried the investigation further and asked whether sounds could be used for individual recognition within the same species. Again they worked with a colony of bicolor damselfish on the coast of Florida. In this colony, five males held contiguous territories, and the call of each one was recorded. The researchers then based the remainder of their experimental protocol on the "dear enemy" effect. This is a phenomenon whereby territory owners learn to overlook the presence of neighbors with whom they have already interacted as opposed to new colony members with whom differences have not been settled yet. So, Myrberg and Riggio observed the behavior of a male while a speaker in the neighboring territory played various calls. If the call was that of the rightful owner of that neighboring territory, the listening male showed little reaction, presumably recognizing that everything was as it should be. But if the call was that of another fish from farther away, a greater reaction was elicited, indicating that the call was perceived as different (see fig. 2.2). In this way, the two ethologists showed that all males in the colony could recognize the acoustical signature of their two nearest neighbors. (Interestingly, the fish also reacted strongly to their own personal call being played from a neighboring territory. Either they could not recognize their own call, or they "knew" it must have come from another fish, since obviously they could not be in two places at once.)[10]

A great number of species are known to produce specific sounds in conjunction with aggressive acts such as chasing, ramming, biting, and

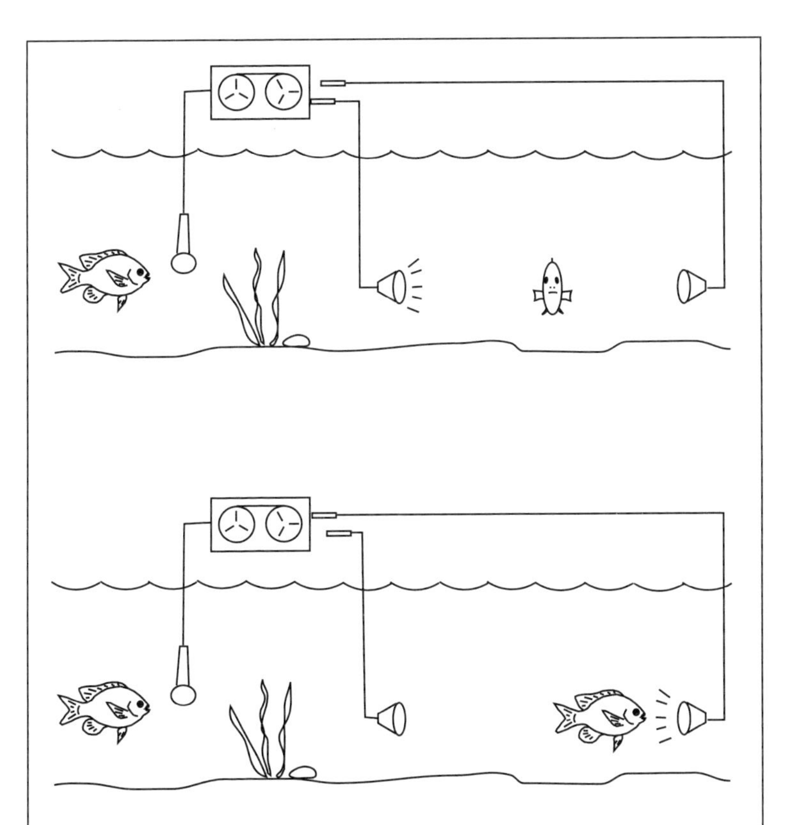

Fig. 2.2. Acoustical recognition and the "dear enemy" effect in damselfish. If, when the territorial call of a neighbor is played back, it comes from the direction of that neighbor's usual territory, little reaction is elicited on the part of a listening male. But if the playback comes from a different direction, the male reacts aggressively.

visual displays that are known to have aggressive meaning.[11] Many of these species are territorial, and the sounds can be heard while the territory owner apparently warns and then expels a trespasser. In the territorial skunk loach, these war cries are so important that experimentally muted residents lose their ability to repel intruders, even though they try to compensate for the lack of sound production by increasing their rate of visual threat displays.

Speaking of visual displays, some of them seem to be specifically associated with concurrent sound emission. For example, some cichlids such as the jewel cichlid shake their heads while producing aggressive sounds. Unfortunately we do not yet know whether headshakes are an aggressive display that enhances the meaning of the acoustical message or simply a by-product of the sound-making action.

An Effective Complaint

The John Dory is an edible fish that is found and commercially exploited in temperate seas. In many languages, it is called Saint Peter's fish (even in English, it is sometimes known as Peter's fish). The origin of this name is related to the black spot the fish bears on each of its flanks. Fishermen in the old days believed those black spots to be the mark of Saint Peter's fingers, as the fisherman apostle was said to have captured the fish but then released it by hand upon hearing its cry of despair, for the John Dory—and this is a fact—grunts loudly when it is lifted out of water.

This legend is charming, although it conveniently disregards the fact that the John Dory does not occur in the Sea of Galilee, where Peter fished. But as an explanation for the origin of this fish's name, the story is as good as any.

Vocalizations can also be overheard during fights for dominance status. Experimental work has confirmed that these vocalizations carry aggressive intent and that they may induce the receivers to think twice about escalating a fight. In one experiment, a satinfin shiner was exposed to its own image in a mirror, and its level of aggression toward this "other" fish was noted. Then, various sounds were played back in the fish's tank. When the sound played was only background noise (the control condition), the levels of aggression in front of the mirror were not diminished. However, when the sounds were those made by a more dominant shiner during a fight, the test fish became more timid and approached its mirror image less often. Interestingly, when the playback featured the sound of a courting shiner, aggression levels remained high. The shiner could therefore tell the difference between the aggressive meaning of a fighting call and the nonaggressive message from a courting call.[12]

Another example comes from the Central American cichlid *Cichlasoma centrarchus*. This cichlid, which normally would direct highly aggressive acts to other fish in nearby tanks, fails to do so when conspecific calls are played back to it. This inhibiting effect does not occur in response to a control sound (in the original study, the control was the playback of "grunts made by a deep-voiced colleague"). Male cichlids also tend to back down more readily in front of a rival that both displays and vocalizes as opposed to one that only displays.[13]

Acoustical intimidation carries more weight when it honestly reflects the fighting ability of a contestant. At the University of Vienna in Austria, Friedrich Ladich observed that in fights between male croaking gouramis (croaking is part of the species name), winners were usually larger than their opponent, and their acoustical signals were correspondingly deeper and louder, although not necessarily more frequent. Although it is theoretically possible even for a small fish to bark a lot, only a large body allows for a deep and loud voice. Therefore, acoustical parameters could be used as honest signals of fighting ability. Contestants could listen carefully to the voice of their opponents to assess relative body size and therefore evaluate their chance of winning a fight, even before the first blow is delivered.[14]

Acoustical signals delivered during a fight need not have aggressive meaning. Some clownfishes apparently make special sounds when they are attacked by dominant individuals. Such whimpers may appease the aggressor, as they do in dogs, although this has not been proven yet.

The sense of hearing is for more than communication. It can also help fishes find food. Numerous studies have shown that free-living sharks, groupers, snappers, and other predatory species are attracted to the sounds of struggling fish. Those sounds are not distress calls by the potential victims but the noise made by their thrashing movements in the water. This noise is characterized by low frequencies and choppiness. For predators, these simple cues are like a bell tolling for dinner. Researchers have in fact succeeded in attracting sharks to nondescript spots in the sea simply by playing back intermittent pulses of low frequency that were artificially generated (see fig. 2.3). Given that these experiments used playbacks only, we know that the sharks did not rely on vision or olfaction because there was no struggling prey to be seen or smelled.[15]

Fishes might also learn to associate the sounds of a feeding shoal with the discovery of a food bonanza. Japanese researchers have recorded the noise made by captive shoals feeding on bait, and after lowering a speaker into the sea and playing back that noise, they saw wild shoals of the yellowtail *Seriola quinqueradiata*, of chub mackerel, and of jack mackerel gathering around the speaker.[16] It is unclear whether the relevant noises were created by the excited swimming motions of the feeding fish or by the chewing action of their jaws. The latter may seem implausible, but clear chewing sounds have in fact been documented in at least two species of fishes, the dwarf seahorse and the lined seahorse.[17] Even divers can sometimes hear the scraping sounds of coral reef grazers. We can easily imagine a scenario whereby shoaling fishes—especially those that live in poorly lit habitats—might eavesdrop on the activity of other shoalmates and rush to join them after hearing them bite into something crunchy, such as a snail or a shrimp. Information sharing is a common characteristic of shoals, and it does not have to be restricted to visual channels. This is an avenue for further research.

Stationary sources of sound could also be used as a beacon for orientation, but so far this has not been proven conclusively. Some experimenters have reported that larval coral fish, when released at night in open waters 1 km away from the nearest reef, unerringly take off in the direction of that reef. Perhaps they use sounds made by crashing waves or by the adult fish already present on the reef as a directional cue.[18]

Sound is also involved in interactions with predators. More than 35 families of fishes include species that suddenly produce sounds

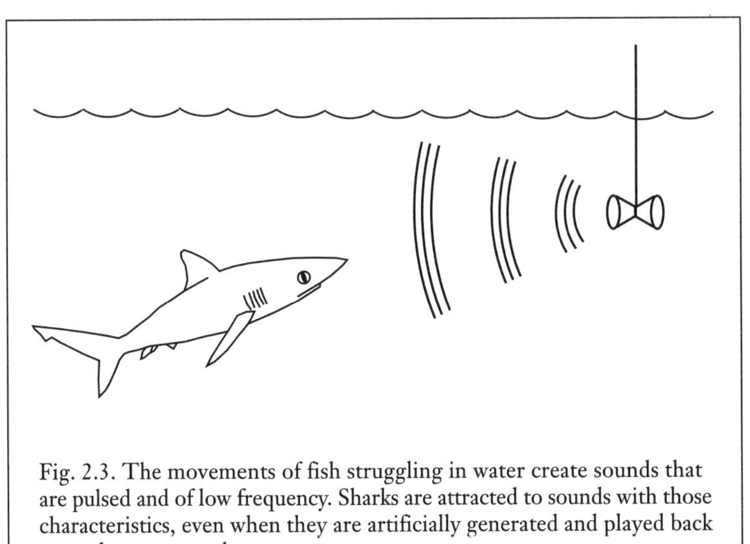

Fig. 2.3. The movements of fish struggling in water create sounds that are pulsed and of low frequency. Sharks are attracted to sounds with those characteristics, even when they are artificially generated and played back on underwater speakers.

when they are disturbed by divers or by other fish. At first, ethologists thought that the function of these vocalizations was to startle predators, in the same way that disturbed butterflies sometimes try to scare their attackers by flashing eyespots on their wings. However, on the few occasions when fishes were overheard giving such calls in response to the approach of a conger eel or a shark, the predator did not seem fazed in the slightest. Then came the idea that the vocalization was a warning signal to other fish. However, seemingly at odds with this warning signal hypothesis is the observation that prey species sometimes *approach* the location of the blazing trumpets. So far there has been no experimental work on this topic.

It was recently demonstrated that the American shad, a member of the herring family, can detect ultrasounds of 25,000–130,000 Hz. Sensitivity in this extremely high range is not as acute as it is around the more normal 200–800 Hz, but it is better than at intermediate frequencies. The authors of the study, David Mann, Zhongmin Lu, and Arthur Popper from the University of Maryland, believe that this exceptional hearing ability is an adaptation by the fish to detect the ultrasonic clicks of their echolocating predator, the dolphin. This reasonable interpretation is supported by observations that the closely related blueback herring often swims away from marine echosounders that emit ultrasound and that ultrasound can be used to keep other herring species away from the water intakes of hydroelectric power plants.[19]

I find it striking that we can find analogies between the acoustical behavior of fishes and that of other animals—mammals, birds, amphibians—whose vocalizations we take for granted far more easily than those of fishes. Contrary to the opinion we might form at first sight, fishes are not deaf and dumb, and we should be aware that sound plays an important role in the life of many species, maybe just as important as in those terrestrial animals that possess powerful vocal cords and large external ears. Fishes may not need a hole in the side of their head, but they still profit greatly from being able to hear.

3

Lateral Line

It brings a sense of wonder to realize that some animals can perceive things that we cannot. Some snakes can pinpoint the position of a mouse just from the heat that radiates from it; some mammals use echolocation because they can produce and detect ultrasounds; some animals—fishes among them—are sensitive to ultraviolet or polarized light. Sometimes these extraordinary sensory abilities are not betrayed by any obvious external organs. Only specialized cells, deep inside the eye or the nose, hint at the animal's superhuman potential. But in some other cases, odd external organs are readily visible and scream for an explanation of what function they could fulfill. Depressions on the head of pit vipers constitute a fine example of this; another is the hump on the head of an echolocating dolphin. And yet another is the lateral line of fishes, readily visible on the flanks of most species, extending from head to tail. What is it good for?

Careful examination under a dissecting microscope reveals that the lateral line is a row of pores. Similar but shorter rows can also be seen on the side of the head (see fig. 3.1). Irrespective of their location, each row of pores connects the outside to a canal that is recessed within the skin, or sometimes within the underlying bone, like a pipe buried in the ground. Because of its multiple connections to the outside, the canal is filled with water, which can slush this way and that way, depending on the pattern of water flow outside the body. Water flow inside the canal bends thin hairs that project from sensory cells within the canal walls. Hair-bending in turn causes the sensory cells to send a message to the brain. Some hairs can bend only along one axis, other hairs along another, and therefore the fish's brain can deduce the direction of water movement from the identity of the sensory cells that happen to be firing signals at any one time. In short, the lateral line enables fishes to perceive the pattern of water flow around their body.

Much less conspicuous than the lateral line or the head rows are a number of free-standing pores, scattered all over the body of fishes, sometimes in loose rows, sometimes in pairs, sometimes by themselves. These pores represent the openings of subcutaneous pits, which also contain sensory cells and hairlike projections. These pores are considered to be part of the same system—called the lateral line system—as are the head rows and the lateral line itself. However, the

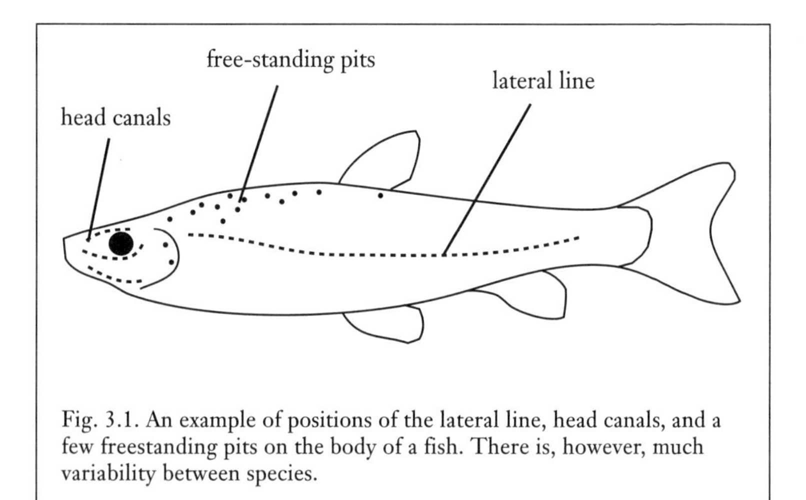

Fig. 3.1. An example of positions of the lateral line, head canals, and a few freestanding pits on the body of a fish. There is, however, much variability between species.

precise function of the pits is not quite the same as that of the canals. It is generally recognized that the pits are mostly responsive to the directionality of water displacement, whereas the canals are sensitive to accelerations in water displacement.

As with the olfactory and acoustic senses, one of the best ways to determine the function of the lateral line system is to knock it out and note how the behavior of the fish is affected. One long-standing method for blocking the entire lateral line system has been to dip the fish for a few hours in a solution of either cobalt ions or the antibiotic streptomycin. Both agents temporarily suppress the proper working of the receptors. Another inventive method, one that blocks the canals and hence the water movement within them while leaving the pits intact, is to stuff the canals with fine mammal hairs inserted through one of the pores. Dipping the fish in a weak solution of the antibiotic gentamycin also destroys the canal receptors while sparing the pits. For the surgically gifted, it is also possible to cut the nerve that links the lateral line to the brain. On the other hand, destroying the pits while leaving the canals intact is harder to do in a systematic way. Fortunately, for most practical purposes the crude expedient of gently scraping the skin where the pits are known to be concentrated seems to work well.

Being able to perceive water flow and minute changes in its speed has many useful applications for fishes. Because of the lateral line system, a fish can feel the presence of nearby objects, can synchronize its swimming motion with that of other fish in a shoal, can detect the direction of running water and maintain position within it, can communicate information to other fish, and can perceive surface waves produced by struggling insects that have fallen in the water.[1]

The mechanism of object detection works like this: When swimming, a fish pushes water in front of itself. At a constant speed in an open area, the resistance offered by the water is constant, and the water flow around the fish is stable. But if the fish approaches a stationary obstacle, that object adds to the resistance of water and disrupts the water flow around the body of the fish. The fish can detect this disruption and use it not only to avoid the obstacle but also to obtain an amazingly accurate picture of its conformation.

Cavefishes have been the most popular subjects of study for this ability. Most cavefish species are blind, which means that they cannot locate stationary objects by sight and that an alternate means of detection is likely to have evolved to a high degree. The amblyopsid *Typhlichthys subterraneus* and the characid *Astyanax mexicanus* (the Mexican cavefish, formerly known as *Anoptichthys jordani*) have each been studied.[2] When a new object is surreptitiously placed within their aquarium, these blind fish eventually come close to it, seemingly as part of their normal patrolling, but just as they are about to collide with the object they suddenly veer away from it and start to swim alongside, at a distance of only a few millimeters. Thereafter, the fish spend a lot of time swimming all around the object, always close but almost never touching it, as if to explore and determine its shape. Later during the day, the route taken during subsequent travels within the tank definitely gives the impression that the fish are aware of the object's shape and location.

Blind cavefishes often swim faster than usual when they are placed in an environment that is new to them. At first, this does not strike us as a smart thing to do. Certainly we would not run around if we were blindfolded and in a room full of objects lying about in unknown places. But what seems like counterintuitive behavior in cavefishes is explained by physical considerations related to water and its interaction with the lateral line. Because of the physics of water flow around a moving body, the faster the body advances, the thinner the flow field is, and the easier it is for the lateral line to detect changes in the flow field. So, blind cavefishes probably swim faster to maximize the amount of information that they can glean from their lateral line system.

A lot of work has been done in the laboratory of C. von Campenhausen, at Johannes Gutenberg University in Mainz, Germany, on the ability of the blind Mexican cavefish to perceive small details in the structure of stationary objects.[3] As is so often the case in studies of sensory physiology, the researchers used a conditioning protocol in their tests. They gave their subjects a choice of two tunnels to swim through to get food. On the wall near the entrance of each tunnel was a grid of small vertical bars (each bar was 1 mm in diameter). The interval between the bars differed between tunnels. Passage through one predetermined tunnel was rewarded with food, whereas passage through the other tunnel was not rewarded or was punished with a weak electrical shock or a jet of water. The fish had to distinguish between these two tunnels based only on the disposition of the grid pattern at the entrance.

By and large the fish succeeded in making the correct choice, but only when they happened to enter the tunnel at an angle that brought them close to the grid pattern marking it. If they came from another direction, their choice was as likely to be wrong as right. This indicated that perception of the grids could only be done at a close distance—again, a few millimeters. However, if the fish swam close to the grid, their power of discrimination was impressive: they could distinguish between bar intervals of 10.0 mm versus 8.75 mm and choose the correct tunnel accordingly. This discrimination of spatial intervals could not be performed by cavefish whose lateral line canals had been blocked.

If a fish can detect the presence of a stationary object while approaching it, presumably there is no reason why it could not also detect an approaching object while stationary. The situation would be akin to a person standing by the side of a road and feeling the blast of air generated by passing trucks. But fish operate on a much finer scale. Experiments by Sven Dijkgraaf, an authority on the sensory physiology of fish, showed that blinded European minnows can detect the passage of a glass filament only 0.25 mm in diameter (the equivalent of a fine sewing thread) from a distance of 10 mm (a bit less than half an inch) from their body. The data were obtained using a minnow that had been conditioned to associate the passage of this filament with the imminent delivery of food. Immediately after sensing the passing thread, the fish often turned and snapped at it, as it would food, and it was fairly accurate in doing so. All this from a blind fish that could only rely on its lateral line system.[4]

Sven Dijkgraaf, Fish Sensory Physiologist

Sven Dijkgraaf, born in The Hague in 1908, became a professor of biology at the University of Groningen. He had considerable skill at microsurgery and in the design of simple and precise instruments, which he applied mostly to the study of fish senses (mostly but not exclusively; he is also recognized as the codiscoverer of echolocation in bats, along with but independent of Donald Griffin). He was a pioneer in the fields of fish acoustics and fish electroreception, but it is in the study of the fish's lateral line that he made his greatest impact. In fact, his influence has been so great that a 1989 book proposed to honor him by coining the term "to sven" as the action of the lateral line system. An eye sees, a nose smells, an ear hears, a tongue tastes, and so a lateral line would sven. The authors penned the following verses:

> Dear frog and fish, or newt and shark
> You needn't worry when it's dark
> You'll escape or dine just fine
> Svenning with your lateral line.

In the poem, we note the words *frog* and *newt;* they reflect the fact that many amphibians also have a lateral line system.

We also see the phrase "escape or dine." It indicates that predator evasion and food finding may be important functions of the lateral line system, although by no means the only ones.[a]

Many fishes that feed on zooplankton can do so not only in daylight but also at night. Some of them, like herring, strike at individual planktonic animals in daylight and switch to filtration feeding under low light conditions. No need for the lateral line here. But others continue to strike at individual prey even at night. In most cases, nocturnal feeding is aided by large eyes that are good at collecting what little light is available. But eyes, no matter how efficient, cannot work in complete darkness, and yet some fishes can still strike at individual prey in such conditions (as revealed by infrared viewing). Perhaps these fishes are using their lateral line system.

Denise Hoekstra and John Janssen at Loyola University in Chicago conducted a study to determine whether fish in fact used their lateral line system to detect prey. After observing that mottled sculpins fed at night in the field, the two biologists brought some of these sculpins to the laboratory and surgically blinded them. A few hours later, these blind fish were behaving normally within their tanks. They were given dead shrimp to eat. If a shrimp came to rest on the bottom and was not moved, the fish did not react to its presence. However, when a dead shrimp was tethered to a nylon line and moved by hand, the fish struck at it. This suggests that detection depended on movement by the prey and not on its odor. When the pores of a fish's lateral line system were covered by paste on one side of the head, the sculpins stopped responding to moving prey on that side, but they still spun toward prey on the other side. This was convincing evidence that the lateral line is involved in prey detection.[5]

Similar results were obtained in another laboratory for the torrentfish *Cheimarrichthys fosteri*. These fish were not blinded but were kept in complete darkness and observed under infrared light. They had no trouble catching insect larvae that drifted in a current. However, after a cobalt treatment that temporarily impaired their lateral line system, the capture rate declined precipitously. Success at capturing prey was only restored progressively as the lateral line system regenerated itself over the next few weeks.[6]

Both sculpins and torrentfish seemed to detect prey most often when it fluttered alongside the head of the fish. In these species at least, it is the head canals more than the lateral line on the flank that discern the water displacement generated by the legs of swimming prey, or by the whole body of the prey if it is big enough.[7]

John Janssen's work comparing the lateral line system of four species of Antarctic fishes further illustrates the relationship between prey position and lateral line anatomy. Two of these species were planktivores with dorsally oriented mouths, indicating that they normally approached their prey from below; in fact, one of these species lived just beneath the pack ice and fed on the plankton that clung to the undersurface. Appropriately, the lateral line system of these fishes was better developed dorsally than ventrally. The other two species were benthivores and therefore fed near the bottom. Understandably, they had more ventrally oriented mouths. The key point is that their lateral line system was also better developed ventrally, at least in the head area.[8] Studies of this kind, called "comparative" because they compare the biology of many species, are useful because they can disclose correlations between ecology and morphology. In the present case, the correlation between prey location and the effective position of the lateral line system shows that the lateral line plays a key role in prey detection.

Even species that feed on other fishes rather than on smaller organisms may use their lateral line to some extent. Under infrared light in otherwise complete darkness, bluegill sunfish have been observed to strike at small goldfish from a distance of 2 cm (this was the attack distance; the detection distance may have been greater). However, after cobalt treatment that inactivated their lateral line system, the sunfish attacked if they made direct contact with the prey. We can conclude that bluegills can feel the presence of other fishes from a distance, albeit only a short one.[9]

There is only scant evidence that the lateral line can help a fish notice the presence of predators. Certainly the goldfish in the previous example did not seem too good at it. The only work I am aware of was done on the larvae of various marine fishes (herring, halibut, cod, plaice, and flounder). At their young age, such larvae do not yet have a lateral line, only shallow pits on their body (the canals develop at a later stage). In darkness but under infrared illumination, researchers have measured the reaction of these larvae to the approach of a glass rod 1 mm in diameter. Many larvae had to be touched directly with the rod before they reacted, but others displayed an escape response as soon as the rod got to within a few millimeters of them. The response was weaker after a streptomycin treatment.[10] Could we extrapolate from glass rod to predators? Could these larvae detect approaching predators from a short distance even in complete darkness? Would this be enough to give them a good chance of hightailing it in time? This seems unlikely to me, mainly because of the very short reaction distance shown by the larvae in this experiment.

Animal Behavior Science and Animal Welfare

So far, I have described experiments in which nerves were surgically cut, olfactory receptors were cauterized, and lateral lines were desensitized. Fish were made to work for their food or placed at risk from predation. Many people find such manipulations distasteful if not downright unethical. Because most scientific research on animals is ultimately funded with taxpayers' money, it is important to consider public opinion about how animals should be treated in scientific laboratories.

It is true that until about twenty years ago, there were no regulations to ensure that animals did not suffer unduly for the sake of research. Nowadays, however, at least in modern countries, no publicly funded scientific experiment can proceed without having first been approved by local (usually university-based) or national animal care committees. Such committees are made up of people from the research and veterinary communities as well as members of the public. Their mandate—which they fulfill conscientiously, at least in my experience—is to prevent animal suffering unless absolutely necessary for a very important goal and to minimize the number of animals used in experiments. They encourage the use of animals that take well to captivity—many fishes are good in this respect—and that can adapt quickly to physiological and psychological disturbances.

Behavioral scientists who study sensory abilities tend to use surgical (and hence, potentially suspect) procedures more often than others do. But like all of their colleagues interested in animal behavior, their main concern at least in these modern times is to ensure that their subjects are treated with respect. Anesthetics are used during operations, and the eventual consequence of such operations should not greatly cripple the animal, or if so only for a very short time.

Water can be displaced not only by predators and prey but also by conspecifics. One interesting application of this fact is that schooling fish can use their lateral line to perceive eddies and turbulence in the wake of their schoolmates and therefore maintain their position within the group. Most species seem to use exclusively visual cues for coordinated schooling behavior; for example, they often cease to move as a unit in complete darkness. However, this may not be true for all species. In one experiment, individual saithe were blinded by fitting opaque caps over their eyes, and then they were placed into a circular tank in which a normal school was already swimming round and round. At first, the blinded fish let the school go by every time it passed, but after an hour or so they started to turn toward the school at each passage, sometimes joining it for short distances. After 3 hours, the blind fish had joined the school permanently. They were changing position more often than their sighted companions, but they nonetheless stayed within the school. That the lateral line was compensating for the fish's lack of vision by allowing them to maintain their position within the school was proven by surgically cutting the nerve of the lateral line in the blinded fish. These doubly deprived fish now ceased to school completely.[11]

The ability to perceive eddies created by other fish probably evolved from the capacity to detect water currents occurring naturally. The capacity of fishes to orient themselves relative to water flow (usually facing upstream) is common knowledge. As we saw in chapter 1, this behavior is called rheotaxis. Early experiments had suggested that the lateral line system was not involved in rheotaxis, but a team of New Zealanders under the direction of John Montgomery at the University of Auckland has recently forced a reassessment of this point of view.

The researchers worked with three species that were already known to use their lateral line for prey detection: the torrentfish, the Mexican cavefish, and the Antarctic fish *Pagothenia borchgrevinki.* Under normal conditions, these fish spontaneously face upstream even in currents of fairly low velocity, as low as 0.5 cm per second. However, after the free-standing pits were destroyed by scraping the skin, or after a treatment with cobalt ions which inactivated both the canals and the pits, the fish no longer oriented in the slow current. The velocity had to be increased eightfold before normal orientation was restored. This result shows that although very strong currents can be detected by the fish (probably through pressure sensors in the skin), lower velocities require the normal functioning of the pits in the lateral line system, at least in these species.[12]

A fish's perception of natural currents can lead to more sophisticated behaviors than just facing upstream. Anglers are familiar with the following habit of trout in a fast stream: the fish take up positions behind rocks, where they face upstream, and they remain there with no contact with the ground and rather few apparent swimming movements. Obviously, these fish can find the one spot behind the rock where eddies and turbulence negate the pull of the main current (fig. 3.2). This behavior, called obstacle entrainment, allows trout to save energy. Experiments with brook trout in swimming tubes equipped with artificial rocks have revealed that this behavior is made possible by vision during the day (the fish can visually maintain its position behind the rock) and by the lateral line at night (the lateral line system informs trout about the fine details of water flow behind the obstacle). We know this because trout whose lateral lines

Fig. 3.2. Fishes can use the lateral line on their flanks to detect water currents. This organ allows trout, for example, to find the sweet spot behind rocks where eddies negate the pull of the overall current so that they can maintain position in a stream without expending too much energy, even at night when they cannot see the rock.

were denervated were still able to remain behind a rock in strong current during the day, but in complete darkness they were swept away; intact fish, however, were able to maintain position in both light and dark.[13]

In some cases, fishes could displace water for the specific purpose of communicating information to others. In fights between cichlids, for example, the two protagonists often position themselves side by side and direct vigorous sweeps of their tail toward each other (see fig. 3.3). Such tail-beating could transmit information about an individual's strength. In the same way that we can judge the strength of an adversary by the size and speed of the rocks he throws at us, a

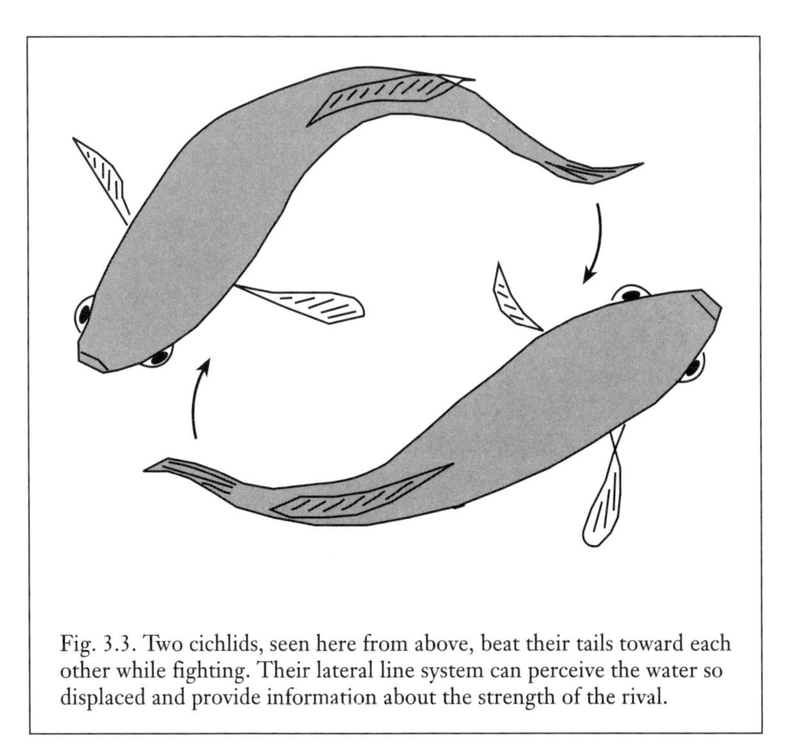

Fig. 3.3. Two cichlids, seen here from above, beat their tails toward each other while fighting. Their lateral line system can perceive the water so displaced and provide information about the strength of the rival.

cichlid can assess the power of an opponent by the volume and velocity of the water it displaces.

Another, more peaceful example can be found in the courtship and spawning activities of many fishes. There, quivers of the whole body often signal readiness to mate, if not the actual act of releasing eggs or sperm, and such quivers undoubtedly displace water. They can therefore be perceived through the lateral line and spur sexual partners into action, as has been reported in salmon.[14]

Another use of the lateral line system, this time limited to head canals, is the perception of surface waves by fishes that feed on drowning insects. An insect struggling on the surface generates waves that present a typical frequency spectrum, different from the waves caused by wind or by falling leaves. Fishes such as the topminnow *Aplocheilus lineatus* and the African butterflyfish *Pantodon buchholzi* can detect and recognize these waves by positioning their head at the surface. If their head canals are intact, these fishes can sense waves that are no bigger than a fine hair. They can also determine the direction of travel of these waves by comparing sensory inputs from each side of the head. (In the preceding chapter, we saw that this mechanism does not work for water-borne sound waves because of their high speed of travel; surface waves, however, move slowly enough.) The fishes can even evaluate the distance of a struggling insect by analyzing the frequency spectrum of the front waves, which changes predictably as the wave train travels farther from its source. With computer-controlled air puffs blown on the surface next to the head of a topminnow, researchers can make the fish move short or long distances simply by creating wave trains that mimic near or far sources.[15]

In Siamese fighting fish, the parental male can communicate danger to its young through surface waves. In their first few days of independence from the air-bubble nest, young fighting fish stay in contact with the surface—like most anabantids, they need to breathe some air. The parental male stays nearby, and if he senses danger, he shakes his pectoral fins close to the surface. The surface waves thus generated are perceived by the young at a distance of up to 40 cm, even in complete darkness. The young then swim in the direction of the source, and this action brings them close to the male, who can then suck them up into his mouth and carry them back to the safety of the nest. Here again, artificial stimuli can be used to fool the young fish and make them gather in the vicinity of points of impact at which appropriately defined air streams have struck the water.[16]

As we have seen, the lateral line system has many uses in fishes. This sensory modality may seem odd to us, but we should remember that it is intimately linked to the aquatic environment in which fishes live. The lateral line works as a sensory organ because of the incompressibility and fluidity of water. This organ is found only in fishes—there are even signs of its existence in the fossils of ancient species—and in some amphibians that also live in water. If you ever need to draw a fish, make it real: add a little line on its flank.

4

Electricity and Magnetism

G*ymnarchus niloticus*, or the aba, is related to a group of animals called the elephantfishes. The aba has a long undulating fin along its back, which allows it to move forward and backward with equal grace. It easily avoids obstacles while swimming forward and also, somewhat surprisingly, while backing up. Does it have eyes behind its head? If so, one would hope they might be in better shape than those in the front, which are degenerate and barely functional. Does the aba have a lateral line on its tail? There is no indication of it. Instead the tail appears naked. So how can these fish sense what is behind them? And given that their eyesight is so poor, how can they know what is in front of them?

Welcome to the alien world of electric location. It is nothing like our own sensory world. In it we find electric organs, that is, modified muscle tissue in which hundreds of disklike cells are piled up in long series on the flank of some fishes. The aba has four such spindles running up each side of its body, from the naked tail to approximately two-thirds of the way toward the head. These structures can generate electric current in the water surrounding the fish. Equally important are electroreceptors. In terms of development, electroreceptors arise from the same embryonic structures as do the hair cells of the ordinary lateral line, and they are innervated by the same nerve as is the lateral line. In a word, they greatly resemble the lateral line.

There are two types of electroreceptors, ampullary and tuberous. The ampullary ones are used in passive electrolocation, that is, the detection of low-frequency electric events generated by all living aquatic organisms as a by-product of their physiology and movements. As we shall see, this is very useful for finding hidden prey from a short distance. Tuberous organs, for their part, are built for active electrolocation, that is, the perception of mid- to high-frequency electric fields generated either by the fish itself, thanks to its special electric organs, or by another fish of the same species. Perceiving changes in one's own field is useful for locating objects from a short distance through the field distortions created by these objects (be they in front or behind the fish), whereas perceiving another fish's electric discharges opens the door for communication.

As far as we know, only cartilaginous fishes (sharks, rays, skates, and ratfishes, as well as the mostly cartilaginous sturgeons and paddlefish), all catfishes (the order Siluriformes), and two African notopterids (*Papyrocranus afer* and the featherback fish *Xenomystus nigri*) are capable of passive electrolocation. They use this sense to find prey buried in sand or mud.[1] No matter how quiet a concealed prey may be, it cannot prevent its heart from beating, its mouth from circulating water through the gills, or its tonic muscles from contracting. These actions are created by electric events within muscles, and this electricity can carry through body, sand, mud, and water. As an analogy, we may think of how electrocardiograms can be recorded from the surface of our own bodies; essentially, the electric activity of the heart muscle is transmitted through the rest of our body and picked up by instruments applied to our skin. Bioelectric events of this kind are weak, but they can still be perceived by properly equipped fishes from a distance of up to 30 cm (approximately 1 foot). Sharks and skates, for example, can detect voltage gradients as small as 0.01 microvolt (that is, ten-billionth of a volt) per centimeter, the greatest electrical sensitivity known in the animal kingdom.

Ad Kalmijn, from the Scripps Institute of Oceanography in La Jolla, California, has elegantly demonstrated the important role of passive electroreception in sharks. He worked with the dogfish shark *Scyliorhnus canicula.* Some of those sharks were kept in a basin with a layer of sand at the bottom, and they were fed with plaice. When cut up pieces of plaice were buried in the sand, the sharks were attracted by the odor given off by those tender morsels but could not find their precise location—dead tissues do not generate bioelectric fields. In contrast, when a live plaice was buried, the sharks found it and quickly dug it up. If the plaice was buried in a chamber of agar that masked all olfactory and visual cues without blocking the electric field, the sharks still found the plaice and tried to attack it in the same manner as before. However, if the agar chamber was also covered with a high-impedance plastic film that masked the electric field, the sharks could not locate the plaice. Better still, if a pair of active electrodes was buried a short distance from a piece of dead fish lying on the bottom, the sharks ignored the meat in plain sight and instead attacked the invisible electrodes. Tests at sea with free-living smooth dogfish and blue shark also showed that sharks spontaneously bite active electrodes while overlooking inactive ones.[2]

In some planktivores, passive electrolocation could even be used to detect prey as small as water fleas (*Daphnia*). This has recently been demonstrated in the freshwater North American paddlefish, an unusual fish blessed with an elongated rostrum in front of its head. The function of that organ has long been cloaked in mystery, but now a team of biologists from the University of Missouri proposes that the rostrum is an electrolocator. In complete darkness except for infrared light, captive paddlefish easily gobbled up individual brine shrimp as the small prey passed within 9 cm of the paddlefish's rostrum, and they caught smaller *Daphnia* at 6 cm. As we have seen previously, this response could be mediated by the lateral line system, but results from two additional experiments argued against this possibility and in favor of electrolocation. In a field of air bubbles meant to confuse the mechanosensory lateral line, the fish were still able to catch brine shrimp. Moreover, when brine shrimp were coated with agar, which immobilizes the prey but is electrically transparent (agar having the same conductivity as water), thus minimizing mechanosensory cues while preserving electric ones, the fish could still catch the shrimp without problem. Empty agar particles, in contrast, were seldom captured because they emitted no electric activity. Because paddlefish were also seen to make feeding strikes at activated electrodes but not at dormant ones, we are left with the conclusion that these fish are passive electrolocators and that they use this sense to find their prey. The huge size reached by these fish (up to 2 m in length and 75 kg in weight, or 6.5 feet and 165 pounds) in the Mississippi River basin bears testimony to the proficiency of their feeding technique.[3]

Some species use electricity in a more proactive fashion while foraging: not content simply to locate prey electrically, they also use their electric organ to generate powerful charges that stun their victims—a technique, known as electrofishing, that has also become common practice among fishery biologists. These shocks can also discourage the unwanted attention of the emitter's own predators. The electric catfish of Africa, electric ray of the Mediterranean, and electric eel of the Amazon and Orinoco River basins have notorious and long-standing reputations in that regard, having painfully zapped quite a few Egyptians, Romans, and native South Americans in ancient times, and many more people along the way ever since.[4] These fishes unleash a pulsed DC current that is said to be particularly efficient for shocking prey and predator alike, but the discharges are not strong enough to kill the prey, even though they may reach 500 volts at 1 amp in the case of the electric eel.

An Electrifying Performance

But Nature has with poison armed her sides
And added cold that in the marrow glides
From which all animation numbness feels
And, through each vein, the chill of winter steals.

Thus did Claudian, a fourth-century Roman poet, describe the electric ray, *Torpedo torpedo,* and the 45-volt shocks it can deliver. The poem also refers to the numbing effect of those shocks, which helps explain the Latin name *torpedo* bestowed on this fish, the root of which is still present in today's language as the adjective *torpid.* The ancient Greeks too were well aware of this characteristic of the electric ray. The great Socrates himself once had the effect of his speeches compared with that of the torpedo fish. His speeches were said to be "torpifying," presumably meaning mind-boggling (or leaving the audience speechless), although his personal enemies might have implied mind-numbing (that is, soporific).

Other fishes produce charges that are weaker and used for figuring out one's surroundings—a behavior called active electrolocation, in contrast to the passive electrolocation described above in which the electric events are generated by other living beings rather than by the locator itself. Chief among these so-called weak electric fishes are the South American knifefishes (gymnotiforms) and the elephantfishes (mormyrids). These animals have weak electric organs that discharge either very abruptly or in a sinusoidal manner, depending on the species. The abrupt discharges are probably the fastest electric phenomenon in all of biology, faster even than the action potential of our neurons. Discharges follow one another almost continuously throughout the life of the fish, but their power level is much too low to be detected by human handlers or to be used as a means to stun prey. The discharges are potent enough, however, to create a stable electric field around the body of the fish.

Electroreceptors monitor the status of this field at the body surface. If the animal comes close to an object and the object does not have the same conductivity as water, then the electric field is warped not only around the object itself but on the surface of the fish as well—as confirmed experimentally by electrodes implanted on the fish's skin (see fig. 4.1). The electroreceptors pick up this change, and the brain analyzes the nature and the location of the alteration to get informa-

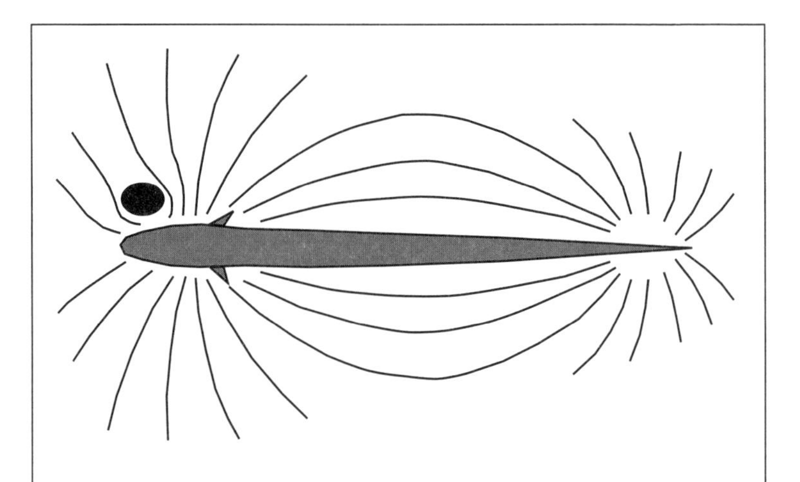

Fig. 4.1. Dorsal view of a weakly electric fish and its electric field. The black object near the head has a poorer conductivity than water. Note how it distorts the field lines at the surface of the fish as compared with the field lines at the side, where no object is present. An object with a better conductivity than water would bend the lines inward toward itself, whereas an object with a conductivity identical to water, like a ball made of agar, for example, would not bend the lines at all and would be electrically transparent, undetectable by the fish's electric sensors.

tion about the object. In this way the fish can feel from a distance. Fishes can even distinguish between close and far objects within that range in complete darkness.[5] It is only a short range (6–7 cm), but active electrolocation remains useful in the muddy waters in which knifefishes and elephantfishes are known to prosper.

An elegant experimental demonstration of active electrolocation would consist of training a weak electric fish to respond to the presence of a small glass rod even when the rod is hidden inside an opaque but porous pot. If the fish could distinguish between a porous pot that contains a glass rod versus one that does not, this result would be evidence of electrolocation. Obviously the discrimination could not be based on visual or mechanosensory (lateral line) information. On the other hand, being porous, the pots would not affect the electric field and would therefore be electrically transparent. They could not hide the glass rods from the electric sense of the fish. It would be as if the pots did not exist. Such experimental evidence has in fact been obtained with our graceful friend the aba, by H. W. Lissmann, the discoverer of electrolocation in fishes.[6]

The weak electric system could also allow a fish to monitor its own position relative to the ground or surrounding rocks. In the knifefish *Eigenmannia*, the electric system mediates the so-called ventral substrate response. This is a habit exhibited by some species—electric or not—in which the fish orients its body more or less perpendicular to the surface of the nearest object, with the belly facing that surface. (Some fishes can therefore be seen swimming upside down underneath ledges or along the top side of caves.) In non-electric fishes, this orientation is visual, but blinded *Eigenmannia* detect the surface with electroreceptors. If a Plexiglas plane is presented at a 45° angle from the vertical, a blind fish swims along it at a 30° angle from the vertical (see fig. 4.2); the incomplete tilt probably indicates that gravity is still perceived and keeps the reaction partially in check. But if the plane is made of agar, which is electrically transparent, then the fish tilts its body only by 13°. Obviously it is less aware of the plane's true angle. The fact that there is still some tilt may mean that the plane is not completely transparent or that it is partially detected through the lateral line.[7]

The weak electric system may also be involved in the exploration of novel environments. For example, blind *Gnathonemus petersii* (they are elephantfishes) can easily find the only aperture that allows them to cross a newly installed partition within their aquarium, even though they cannot see it with their eyes. Their electric sense must be implicated because when these blind individuals are rendered electrically silent (that is, after denervation of their electric organ), they can no longer find the opening. Conversely, apertures that are obstructed by electrically transparent agar are still found by weakly electric fish, and

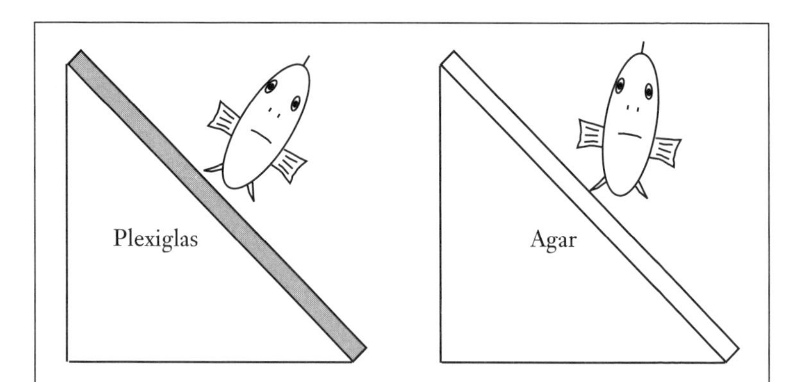

Fig. 4.2. Ventral substrate response in a weakly electric fish, showing that active electrolocation is used to detect the orientation of the substrate. If the inclined plane is made of Plexiglas, which distorts the field generated by the electric organ, the fish swims almost perpendicular to the plane, a behavior that is characteristic of the ventral substrate response. But if the plane is made of agar, which has the same conductivity as water and is therefore electrically transparent, the fish orients its body noticeably closer to the vertical.

these fish keep bumping into the agar barrier in their electric but mistaken notion that the aperture is open and passable.[8]

The presence of the weakly electric system was discovered in the 1950s, and right away its importance for active electrolocation was emphasized. It was not until the 1970s that a second role, as important if not more so, grabbed the imagination of biologists—the possibility of electrical communication between individuals. Communication is possible because the rate and waveform of the electric discharges can vary between species, between sexes, between individuals, or even between situations in the same individual. Moreover, some fish can temporarily interrupt their normally continuous train of discharges, and these pregnant pauses can be full of meaning. The effective range of communication by electric signals can reach a little over 1 m depending on water resistance.

In terms of functions, electric communication is strikingly similar to acoustical vocalization.[9] Some of these functions are concerned with reproductive activity. In some species, males switch to new electric calls during courtship, resuming their regular programming only after the mating season is over. In species in which each sex has its own distinctive pattern of discharges, females are attracted to the pattern of males, and males to the pattern of females. Females can even be induced to spawn in the vicinity of electrodes that imitate a male signal (the "spark" of love).[10] As expected, through natural selection, both males and females prefer the electric pattern of their own species to that of other species.

Other functions relate to aggression. Aggressive individuals often precede their attacks with an increase in discharge rate, whereas submissives may stop emitting altogether. This submissive behavior seems to work; researchers have found that individuals rendered electrically silent (again through denervation of their electric organs) are seldom attacked by dominant fish.[11] Finally, individual recognition can also be based on electric signatures. In *Gymnotus carapo*, territory neighbors recognize each other through individually distinctive discharge waveforms, as demonstrated through playbacks and the "dear enemy" effect[12] (this experimental protocol was described in the chapter on hearing).

The fact that weak electric fishes can use their electric sense to communicate with one another leads to an interesting question: given that these fish continuously discharge their electric organ throughout their life, how can a fish distinguish between its own electric bursts and those from another fish? In elephantfishes, the problem is solved by the presence of two types of electroreceptors. One of these two types is automatically and briefly shut down each time the fish discharges. Therefore, any signal picked up by these electroreceptors has to come from another animal. Elephantfishes also have the habit of "echoing" the discharges of other individuals. They discharge their own electric organ a fixed time after sensing the electric signal of another fish. This response time is extremely short—approximately 12 milliseconds—probably the most rapid form of communication in the animal kingdom. The significance of this value is that very few species intersperse their own signals by 12 milliseconds, and so there is a window of silence 12 milliseconds after each discharge from a stranger.

Knifefishes also display a peculiar behavior called the jamming avoidance response.[13] Somehow these fishes can keep track of the discharge rate of an interlocutor while remaining aware of their own. If the two rates are too close, each fish alters its frequency of discharge so as to widen the gap between the two. In a sense, they do not want to get their wires crossed. In the laboratory, it is possible, using artificial signals, to force a knifefish to decrease its frequency of firing just by exposing it to a high but slowly decreasing signal rate (or an approach from above) or to increase its frequency of firing by switching to a low but slowly rising signal rate (an approach from below) (see fig. 4.3).

Some fishes are also sensitive to magnetic fields. From a physicist's point of view, electric and magnetic fields are conceptually related.

But in terms of fish physiology, the sensory organs involved in their detection are quite different. Whereas electrosensors are located in the skin, magnetosensors seem to sit inside the head. Magnetite particles, Fe_3O_4, which are thought to form the core of all known biological magnetosensors, have been found in the head of salmon. Particles that are likely to be magnetite have also been discovered inside specialized cells within the nose of trout. These cells were found to be connected to the brain by a special nerve, and signals were detected on this nerve when the fish perceived magnetic anomalies in their environment. Fishes in fact represent the first vertebrates in which such potential magnetoreceptors *and* associated neurological hardware have been identified.[14]

The only reason an animal might want to perceive the earth's magnetic field is to integrate this information within a compass mechanism

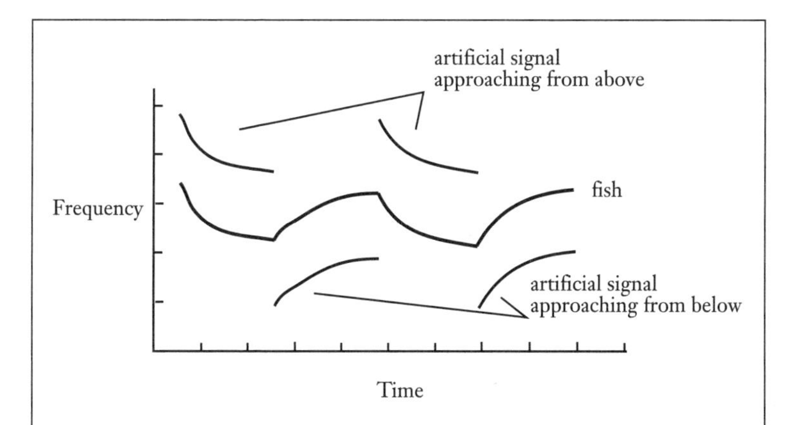

Fig. 4.3. To avoid getting their wires crossed, neighboring electric fish try to keep the frequency of their electric discharges apart. This is called the jamming avoidance response. It can be demonstrated in the laboratory by exposing an electric fish to an artificial signal and noting how the fish tries to maintain a steady difference between the frequency of its discharges and that of the signal.

that will help it to find its way around, much like modern backpackers do with manufactured compasses. For example, if a fish used to living in a small home range at the eastern end of a lake is chased away by a predator, to find its way back it only needs to swim in an easterly direction. A properly calibrated magnetic compass could help it achieve this.[15]

Experimental protocols for demonstrating the existence of a magnetic compass in fishes are simple enough, although they necessitate the use of special equipment capable of creating artificial magnetic fields or modifying the natural one. Once so equipped, all that is left to do is to obtain fish that show a consistent orientation in a magnetic field—either naturally or because they have been trained to do so—and then alter the field. One expectation is that the initial orientation will disappear if the magnetic field is abolished. And if the magnetic field is experimentally rotated, the prediction is that the preferred orientation of the fish will also rotate. In many such tests, the predictions have in fact been upheld.

Lincoln Chew and Grant Brown, at the University of Lethbridge, Alberta, put commercially bred rainbow trout in circular arenas and found that these fish repeatedly—but unexplainably—faced magnetic north. The trout were not lining up by using spatial cues from outside the pool because visible surroundings were regularly moved between trials, and yet the fish still faced north. However, when the experiment was repeated in a room enclosed in mu-metal (a special nickel-silver alloy) and lined with grounded copper strips to neutralize the natural magnetic field, the orientation of the trout became random.[16] Similar results have been reported in leopard sharks, which also, intriguingly, oriented in a northerly direction when placed in a normal magnetic field.[17]

At the University of Auckland in New Zealand, P. B. Taylor kept chinook salmon in a rectangular tank oriented east–west, with the water flow and the food coming from the west. The fish, understandably, faced west. Things got more interesting when the salmon were moved, 18 months later, to a circular tank in another location without a view of the sky but within a normal geomagnetic field. Even though the water flow was no longer directional, the fish still faced west. Taylor then installed Helmholtz coils (circular wraps of copper wire energized by a DC power supply) around the arena and, by turning on the power in the appropriate coils, eliminated any trace of a magnetic field. The orientation of the fish became random. In other trials, the coils were realigned to preserve the local magnetic field, but they were rotated 90° clockwise. The orientation of the fish also changed by 90°, but the orientation was either clockwise or counterclockwise: they faced north or south instead of west.[18]

Because the salmon in the shifted field faced either north or south rather than north only, their behavior was called alignment rather than orientation. There is no solid explanation for this quirk of behavior. Perhaps alignment results from an imperfect detection of the magnetic field polarity. At any rate, alignment seems to be peculiar to magnetic orientation. It is not seen in other types of compass, such as visual ones based on sun position.

This is illustrated by the work of Thomas Quinn and Ernest Brannon from the University of Seattle, who studied the migration behavior of juvenile sockeye salmon in Babine Lake, British Columbia. This lake is 150 km long and lies along a northwest–southeast axis. At the time of their capture, the young smolt studied by Quinn and Brannon were intent on leaving the lake on their seaward journey. When tested in circular arenas, they oriented properly in one direction: toward the outlet of the lake, which was northwest. This behavior occurred only when they had a clear view of the sky, however, and at such times their performance was not affected by the orientation of the magnetic field (which was normal or rotated 90° counterclockwise). However, when opaque covers were draped over the arenas to mask the view of the sky, the fish started to orient both toward and straight away from the outlet, and they readjusted this alignment accordingly when the magnetic field was rotated.[19]

These findings reveal a hierarchy of compasses, with a sun-based compass (see chapter 6 on telling time) having priority over the magnetic one. When the sun is visible, only the sun compass is used, and proper orientation without alignment is achieved. When the position of the sun cannot be established, the fish is forced to rely on the magnetic compass, and alignment rather than complete orientation is the result. This is unfortunate for those fish that end up facing away from the outlet of the lake, but perhaps they can eventually use other cues, such as water currents, to realize the error of their ways.

Much of what an animal does is in reaction to a stimulus from outside the body. But for a stimulus to initiate a response, it must first be perceived. In the mind of most people, the concept of sensory perception by fishes is somewhat perplexing. This is attributable to sensory chauvinism. For us, water is a foreign medium in which to live, and we have trouble imagining what odors, sounds, and movement "feel" like underwater. And of course we have no idea what electric and magnetic fields feel like. (In fact, aquarists working with electrical equipment near their aquarium full of water must take great care not to subject themselves to any electric current.) But as we have seen in the first part of this book, fishes fortunately are not like humans, and they are supremely well adapted to their peculiar environment, possessing all the necessary sensory abilities.

Once a simple stimulus is perceived, the fish may react in very crude, almost robotic, ways. But a fish is also capable of integrating information from many stimuli, processing this information and then developing fairly sophisticated behavior as a consequence. Already we have seen some examples, with fish performing feats of orientation, spatial memory, and individual recognition. There is more to come in the remainder of the book.

Part 2

Cognitive Abilities

5

Learning

Learning can be summarily defined as the ability to change one's behavior as the result of experience. For as long as scientists have been interested in animal behavior, learning has been one of their major topics of study. Traditionally the subject has been approached from two different angles. In one type of study, comparative psychologists have tried to determine what the animal brain is capable of, that is, what are the limits of an animal's ability to learn. In their experiments they have used carefully controlled conditions and ingenious protocols, usually in the laboratory and with a few key species such as the rat or the pigeon. (Psychologists, by definition, concentrate on the intricacies of human behavior, but comparative psychologists study animals, ostensibly with a view toward comparing animal and human behavior in a way that will help them better understand people, although nowadays this original goal is only paid lip service.) In another type of study, ethologists and behavioral ecologists have examined learning as part of the natural behavior that animals display, or could potentially display, in the wild. They have worked both in the laboratory and in the field, and have looked at a greater variety of animals than comparative psychologists have.

In the first half of the twentieth century, there was some friction between these two schools of thought, psychologists accusing ethologists of being too quick to conclude that learning was absent or present in an animal's behavior before all other alternatives could be discarded through carefully controlled experiments, and ethologists accusing psychologists of studying biologically irrelevant learning tasks. Fortunately, in the past few decades, a more cordial relationship has developed between the two camps, and their methodology has converged. Comparative psychologists now study a greater variety of species and try to relate their findings to the ecology of the animal, whereas ethologists often use more rigorous protocols and controlled conditions to determine the role that learning can or cannot play in the behavior of wild animals.

Fishes have not escaped the attention of researchers in this field. At first, one might think that fishes, being the "lower" vertebrates that they are, would not be as "intelligent" as rats or pigeons and would not measure up to their learning capacity. This turns out to be untrue. For almost every feat of learning displayed by a mammal or a bird, one can find a similar example in fishes. Admittedly, there are some learning tasks that rats can do and that fishes cannot, but these tend to be abstract ones (that is, of a higher order, to adopt the proper technical term) with little apparent relevance for the animal in nature.

Equating intelligence with learning ability may be a wrong premise from which to start anyway. Intelligence and insight are hard to define operationally, but at any rate the yardsticks that are now used to evaluate such cognitive entities do not include simple learning tasks.

Learning plays an important role in all aspects of a fish's life, and it can take various forms.[1] It is therefore possible to classify learning into many different categories. Let's begin with two basic types that have long been recognized by students of animal learning, especially those that come from a psychological background: Pavlovian (also called classical) conditioning and instrumental (also called operant) conditioning.

Pavlovian conditioning occurs when a subject learns that one particular event is usually followed by another. The classical example comes from the pioneering work of the Russian physiologist Ivan Pavlov on dogs. When hungry dogs see food, they start to salivate. If, day after day, the sound of a bell is given a few minutes before the arrival of food, the dogs soon learn that the bell announces food. We know this because the dogs now start to salivate as soon as they hear the bell, well before the food arrives.

Fishes are capable of Pavlovian conditioning. When we approach our fish tanks with a food container in our hands, the fish see us, and they often start swimming frantically near the front glass. Over time, they have learned to associate the approach of a person holding a food container with the imminent delivery of food, and they express this learning with anticipatory movements well before the food itself actually appears.

By and large, food is the most common reward used in studies of Pavlovian conditioning. The fish can show their anticipation in various ways—but they do not salivate. They can increase swimming activity, as in the previous example, or they may simply come out of hiding. Some species may also direct food-procuring behavior at the stimulus that announces food. For example, one study on goldfish used underwater lights that were turned on to signal food delivery, and after a while the fish started to nip at the lights as soon as they came on, in the same way that they would nip at the food itself. Another interesting example comes from a study on the archerfish *Toxotes chatareus*. In nature, archerfish get food by squirting water onto insects that sit on plants above the water, thus knocking the bugs off their perch and into the water, where the fish can grab them. In the laboratory, if a light above the water is regularly turned on a few moments before the delivery of a fruit fly at the surface, archerfish soon start to squirt water at the light itself when it comes on (see fig. 5.1). Psychologists have used the term *autoshaping* to describe this habit of directing reward-related responses at the signaling stimulus.[2]

Researchers who work with male Siamese fighting fish and blue gouramis have used exposure to a mirror or to another male as a reward, not in the sense that the sight of a rival is gratifying for a fish

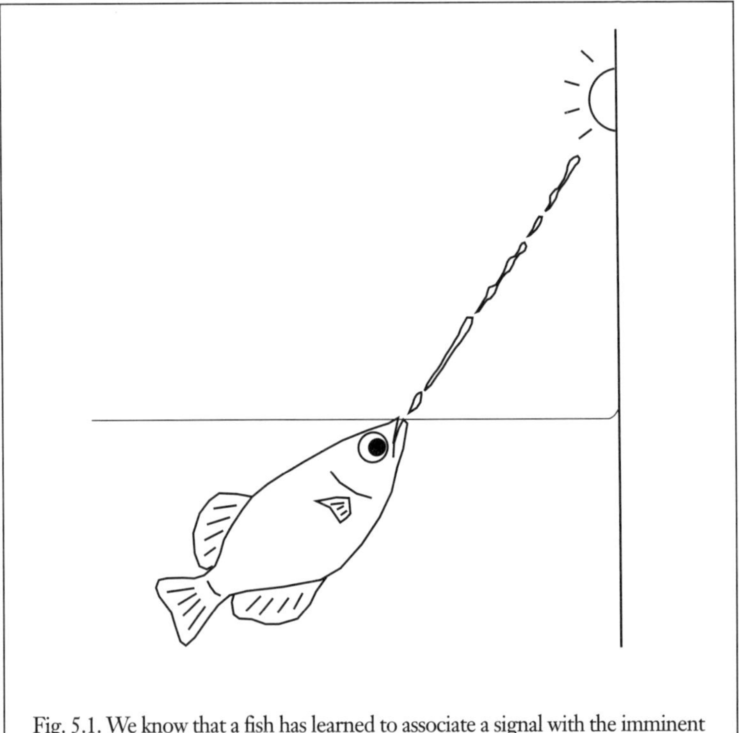

Fig. 5.1. We know that a fish has learned to associate a signal with the imminent arrival of food when it starts to behave toward the signal the same way it would toward food, a phenomenon called autoshaping. Here an archerfish, which normally spits jets of water at insects sitting on plants in the hope of knocking them down into the water, lets fly a volley at a light that has just been turned on to signal the upcoming delivery of fruit flies at the surface.

but rather that sighting a rival will predictably release territorial and defensive behavior in the fish, and this behavior can be paired with a signal such as a red light.[3] Here too autoshaping can be seen: the males often give aggressive displays to the red light itself once they know that it precedes the appearance of a rival.[4]

Happily for the cause of reproduction, access to a ripe female is also viewed as a reward by males. In a series of experiments with the frillfin goby, gravid females were repeatedly dropped from a beaker into the tank of a male. After about a dozen such occurrences, the male learned that the beaker was a signal for the arrival of a female: it started to court the beaker as soon as it appeared above the surface, even if the beaker was empty.[5]

In blue gouramis, territorial males are very aggressive, even toward females at first. But if a male has learned that a red light reliably precedes the arrival of a female, and he is presented with that signal followed with the female a few seconds later, he exhibits less aggression and more courtship toward that female, and eventually fathers more young than males that have not been similarly forewarned. In the same vein, a red light that reliably precedes the entrance of a rival male leads to forewarned individuals defeating their adversaries more often than we would expect by chance.[6] Of course, red lights cannot be found in nature, but we can imagine other signals that could work just as well—an odor or a noise associated with lurking males or females, for example.

Back in the laboratory, signals can be paired to punishment rather than reward. The most common punishment used in research on fishes is a mild electric shock. The most common behavior in anticipation of a shock is freezing on the spot or retreating into a shelter. Another response is physiological: gill ventilation rate and heart rate drop precipitously when fishes expect a punishment. This physiological response is very reliable and has been observed in most fish species studied to date.

Punishment could take more natural forms than electric shocks. It could be exposure to attacks by another individual. Imagine an experiment in which a fish would be placed in the presence of a rival, and this rival had the nasty habit of quickly following all of its frontal displays with a charge and a bite. Conceivably, the victim could learn to associate frontal displays with unpleasant bites. In future encounters, even with less belligerent adversaries, that fish might tend to retreat in response to the frontal display alone.

As we have seen in previous chapters, Pavlovian conditioning is often used as a tool to determine the sensory ability of fishes. Here is one more example involving a system I did not discuss, the eyes. Suppose we want to know if fishes can see in color. We can set up a routine in which two lights, one red and one green, can be lit as a signal. The red light is always followed by food delivery but the green one is not. If the fish start to display a food-anticipatory response when the red light comes on but not when the green one is lit, then obviously the fish can tell the difference between the two lights. If we were careful to choose lights of the same intensity, and if we regularly alternated the position of the green and red lights, then the only cue left for telling the two lights apart would be color. Therefore, we would know that these fish can differentiate between red and green.

The other type of conditioning is called instrumental or operant. Here, instead of the animal receiving an external signal, it is an action by the animal itself that predicts a reward or a punishment. Essentially, the animal learns the consequences of its own acts. For example, a fish can learn that pushing a lever results in the delivery of food. To train a fish to push a lever for food, one needs to proceed in stages. At first, food is dropped into the aquarium every time the fish gets close to the lever. The fish soon starts to spend a lot of time near that beneficent object. Then food is dropped only when the fish actually touches the lever—which it does only by accident at first—and then only when the fish actively pushes on the lever; then, if one so wishes, only when the lever is pushed many times in a row. In this way, fishes can be trained to work for their own food, or to get food on demand. Demand feeding could have practical value in the world of aquaculture, because food could be delivered automatically at times when it was sure to be eaten by the customer fish and not wasted. Demand feeding can also be used by ethologists to record the daily feeding patterns of fishes.[7]

Here again, different types of rewards can be given, and different responses (actions) can be taught. Instead of food, rewards may consist of automatic increases in water temperature and dissolved oxygen, visual exposure to a companion, disappearance of a rival, or even access to the model of a cleaner fish.[8] To obtain these rewards, fishes can be trained to touch or push various objects or to swim along predetermined pathways. There are, however, some limits to what a fish can do in instrumental learning. These constraints are often imposed by hard-wired (innate) behavioral patterns related to the ecology of the animal. For example, it has proven very difficult to train male sticklebacks to bite a rod when access to a gravid female was the reward. In male sticklebacks, the sight of a ripe female instinctively inhibits biting—it releases courtship activity instead—and therefore sticklebacks just cannot learn to bite to get access to a female.[9]

It is not that male sticklebacks are stupid. For example, they can easily be trained to swim through a ring to get access to a female. (Literally, male sticklebacks will jump through hoops to get close to females.) They can also learn many other kinds of instrumental routines that involve more natural behaviors. For example, if a territorial male stickleback is exposed to the sight of another male in a neighboring aquarium, he will go through an extensive repertoire of threat displays, one of which is the "head-down" posture. If the view of the rival is blocked for 30 seconds with a sliding door every time the male gives a head-down display, then in subsequent encounters the fish will start to use the head-down display sooner and more often. The male will have learned that his head-down display is particularly effective in making rivals go away. The reward (the retreat of a rival) is appropriate for the learned response (the head-down threat display).[10]

Rewards can also be the omission of a punishment. In avoidance learning, a fish can be trained to move over a baffle and into the other half of the aquarium when a light goes on, otherwise it will receive a mild electric shock. Pavlovian and instrumental learning are both represented here. The fish must first learn that the light reliably precedes the shock (Pavlovian conditioning) and also that its own act of moving away prevents the delivery of the shock (instrumental conditioning). Again though, some constraints may apply. Take the example of a study on the comparative avoidance learning ability of various freshwater fishes. Striped bass, common carp, and channel catfish learned how to avoid the shock fairly quickly (they had a success rate of more than 50% correct escapes after only 12 exposures). Bluegill sunfish and northern pike were slower, whereas yellow perch and redbelly tilapia just could not be trained successfully. Now, perch and tilapia have an innate tendency to react to danger by remaining motionless and counting on their camouflage patterns, whereas bass and carp tend to flee from danger in nature. So it is not surprising that the former two species—the hiders—did not perform as well as the latter two—the escapers—in a learning task that required movement in response to a threat.[11]

Fishes can also learn *not* to bother performing a given act. Consider the following experiment, which was conducted in the early 1900s. A perch was placed in a tank with minnows, its normal prey, but was separated from them by a transparent partition. At first, the perch repeatedly crashed into the partition in its vain attempts to catch the minnows, but soon it learned to leave these apparently invulnerable prey well alone. It did not chase them even when the partition was removed and the minnows started to swim all over the tank (see fig. 5.2). The perch had learned that for some reason, it was futile to attack the minnows.[12] (Do not try this at home with your favorite incompatible species. Sooner or later, the predator will try its luck again at a passing prey and will quickly realize that they have lost their aura of invincibility. The predator will "unlearn" its previous experience, or to use the technical expression, the learned behavior will become extinguished.)

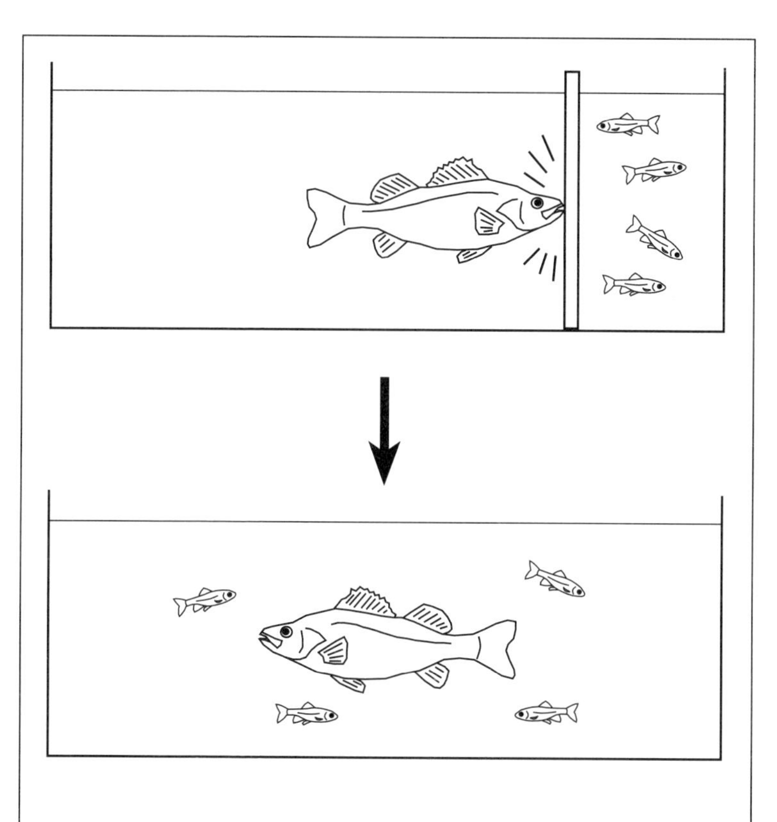

Fig. 5.2. After repeatedly crashing into a transparent partition while trying to catch minnows, a perch gets the idea that these minnows are invincible. Later on, if the partition is removed, the perch leaves the minnows in peace—at first, anyway.

Some types of learning do not fit neatly into the two categories of conditioning. Some of them are covered in the next two chapters. Hereafter are a few more.

Being part of a group may allow an animal to learn how to perform a behavior simply by watching another individual in action. This is sometimes called observational learning or cultural transmission, and fishes are definitely capable of it. The most commonly cited example of observational learning stems from a field study by two researchers, Gene Helfman and Eric Schultz, on a coral reef species, the French grunt. At twilight, juvenile French grunts follow traditional migration routes between their daytime resting sites and their nighttime foraging areas. These routes can be up to 1 km (or 0.6 mile) long. If groups of 10–20 individuals are marked and then transplanted to new populations, they follow the residents along what is for them—the transplants—a new migration route. If the residents are then removed 2 days later, the transplanted grunts, now on their own, continue to use the new route, as well as the resting and foraging sites at both ends. Two days was all it took for the transplanted fish to learn the locations of these sites from the other fish, as well as the fairly long migration route between them. In a control experiment, fish were transplanted to a site from which the residents had already been removed, and they showed no particular directionality of movement at twilight, indicating that the migration route was not obvious and could only be learned socially.[13]

Fishes can also learn the location of good food spots through cultural transmission. If an aquarium is divided in two by a transparent partition and on one side we set up three food patches, of low, medium, and high profitability (depending on the number of food flakes we bury in the gravel), and then we release goldfish on both sides, we will see that the fish on the food side will look around for food and eventually concentrate their efforts on the high-yield patch, while the fish on the other side watch intently (well, they may also try to force their way through like crazy). Then we remove the diners and the food, we wait a while, and we lift the partition. The watchers, now on their own, will probably go straight to where the good food patch was.

Through social learning, fishes might learn not only where to get food but also how to get it. Michel Anthouard from the Université Louis Pasteur in Strasbourg trained juvenile sea bass to push a lever to obtain food. Some fish became proficient at this task, whereas others did not. Anthouard then allowed groups of four naive individuals (the observers) to watch either good demonstrators or bad ones from behind a glass partition (good or bad being defined by how commonly the fish used the lever). Every time the demonstrators pushed a lever, both the demonstrators and the observers got food. After a few days, he removed the lever from the demonstrators' tank and placed it with the observers, and noted how they dealt with the lever on their own. Those observers that had been exposed to the good demonstrators started to press the lever, and therefore obtained food, sooner and more often than did the fish that had been stuck with dumb tutors.[14] Admittedly such results may be hard to extrapolate to natural situations—there are no food levers in the wild—but conceivably various natural foraging patterns such as leaf turning, nosing around rocks, and flushing prey could be learned socially in some species.

The ability to learn the spatial layout of one's environment represents another area of study.[15] We have already seen that salmon can use memory to recognize their home stream or even their home range within that stream. The learned cue was stream odor in that case, but fishes can also use visual landmarks. In the laboratory, fish can be taught to go to specific places that have been marked with Lego blocks or colored discs.[16] Fishes can also learn, either visually or olfactorily, to do the opposite, that is, to avoid bad neighborhoods. For example, it is possible to expose minnows to an alarm substance (see chapter 1) along with water from a specific habitat (a weeded patch in a lake, for example), and thereafter the minnows will display an alarm reaction to water from the weeded area alone. Through association with an alarm substance, the minnows will have learned that the smell of a particular weed patch means trouble.[17]

Spatial memory comes in handy for the small frillfin goby. At low tide, these fish seem to be prisoners of their home tide pool, but when they are chased by mad scientists, they can jump out of their pool and "land" with amazing accuracy in adjacent pools rather than on rocks. Sometimes they jump from pool to pool until they reach open water, a trip that may require up to six different jumps, not all of them in the same direction. This works only when the fish have had a chance to explore the whole area at high tide, when all pools are covered by water and swimming between them is possible. When introduced into an unfamiliar pool at low tide, gobies either refused to jump or landed wrongly on rocks. But after only one night of exploring the new pool at high tide, the jumping behavior became accurate again (see fig. 5.3). The most likely hypothesis to explain this fantastic ability is that the shape of each pool is memorized and serves as the main cue for proper orientation toward the next landing place. Memory of such information is long-lived: gobies tested in the same pools 40 days later still jumped in the right direction. These observations and experiments came from Lester Aronson of the American Museum of Natural History in New York.[18]

Fishes may also learn the order of a series of landmarks and establish routes along them. Like the French grunts mentioned earlier, many coral reef fishes move off the reef on their way to feeding areas at dawn and back along the same route at dusk. Every day, the same route is used. Divers who have observed this behavior could not help but reflect that the fish were following a series of specific landmarks. Supporting this idea, individual fish that were experimentally displaced off the route seemed to move randomly at first, but as soon as they happened to cross their well-known path, they turned and went straight home along it.[19] Unfortunately, observations of this kind do not provide information about the exact nature of the specific landmarks. Experimental removal of putative landmarks would have to be done, although such a task would be made difficult by the occasional absence, to human eyes, of potential signposts, or by the fact that relatively large areas of sand or rock might be used as cues.

Feeding is another activity that is very amenable to the influence of learning. Both the choice of food items (that is, learning about) and the efficiency with which food is handled (that is, learning how) can be improved through experience. In aversion learning, for example, fishes can learn to shun unpalatable prey. One experimenter demonstrated this in a captive group of 150 gray snappers, a predatory coral reef fish. The snappers were fed dead sardines, only some of which were dyed red. The snappers did not discriminate, however, gobbling

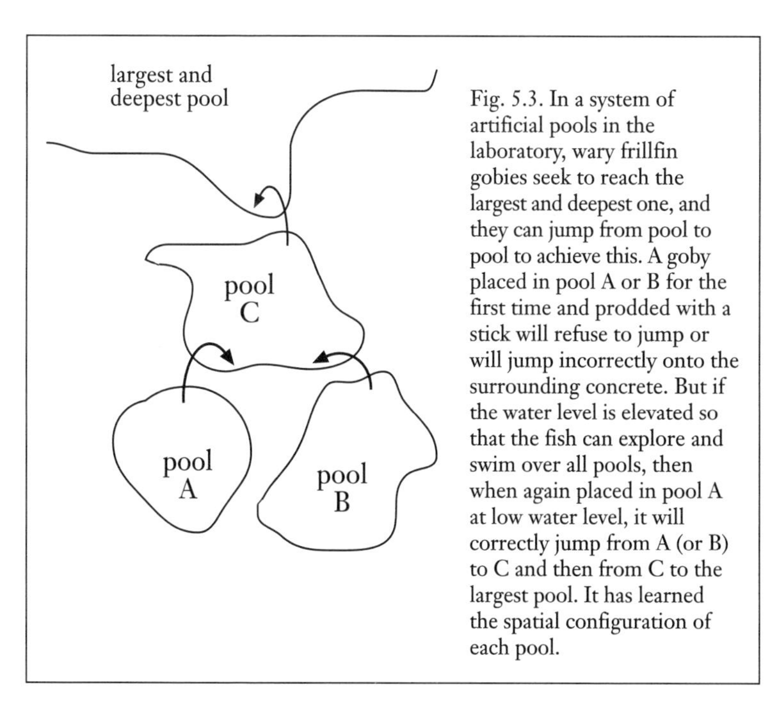

Fig. 5.3. In a system of artificial pools in the laboratory, wary frillfin gobies seek to reach the largest and deepest one, and they can jump from pool to pool to achieve this. A goby placed in pool A or B for the first time and prodded with a stick will refuse to jump or will jump incorrectly onto the surrounding concrete. But if the water level is elevated so that the fish can explore and swim over all pools, then when again placed in pool A at low water level, it will correctly jump from A (or B) to C and then from C to the largest pool. It has learned the spatial configuration of each pool.

up both kinds with equal gusto. Then the red sardines were rendered distasteful by the simple expedient of sewing medusa tentacles into their mouths. The whole group of snappers quickly learned to avoid the red sardines. Twenty days later, they still would not touch those red devils, even though the experimenter had long ago stopped stuffing them with tentacles.[20]

A more natural, if less entertaining example comes from the eating habits of largemouth bass. These fierce predators can quickly learn to avoid toad tadpoles while still consuming frog ones.[21] Toad tadpoles are notoriously distasteful and, like the eggs or the adults, nauseate those predators naive enough to eat them. (It must be said that a very hungry bass, its back to the wall, will stoop to eat toad tadpoles. But if the reaction of other fishes that mistakenly take tadpoles in their mouth is anything to go by—they violently shake their head and you can almost see the grimace on their face—having tadpoles on the menu is no great culinary experience for a fish.)

Fishes that are offered a novel food item will often ignore it at first. Eventually, however, they will sample this new food, and if they find it palatable they will start eating it more regularly. If the food is live prey, the skill required to catch it and to handle it properly can then be improved through learning. The fish become capable of detecting the prey—orienting toward it—at a greater distance, initiating the attack on it sooner, succeeding at catching it more often, and wasting less time to position the prey correctly within the mouth before swallowing it. These improvements are realized more quickly if the fish are hungry or if only one type of prey is offered. If two types of prey are presented alternately, the fish do not learn as quickly as when pure diets are used.[22]

Memory for these skills, however, is limited. If a stickleback that has learned how to forage for prey A is then switched to a diet of prey B for 10 days, and then back to prey A, it will handle prey A as naively as it did before the original learning. If its foraging skills are not regularly reinforced, a stickleback forgets them within 1–4 weeks. There is some evidence that sticklebacks that come from habitats in which prey populations are unstable forget their foraging skills more easily. By forgetting outdated skills more quickly after a given type of prey has disappeared from the environment, these fish may be able to concentrate on learning new skills more thoroughly.[23]

A fish can hone its foraging skills not only as a function of prey type but also according to habitat type. For example, bluegill sunfish are known to develop a fast searching speed when kept in open water and a moderate speed when kept in vegetated areas. These strategies maximize the chances of prey detection when properly used in their respective habitats. As in the case of switches between prey types, regular switches between habitat types slow down the learning process and may lead to poorer foraging success.[24]

To avoid becoming food is as important as getting food, so learning can also apply to the avoidance of predators. As in food learning, fish can both "learn about" and "learn how." Learning about means recognizing predatory species. Some fishes have an instinctive knowledge of what their predators look like, but others do not. In this latter case, recognition occurs only after a fish or one of its close neighbors has experienced a predatory attack. Many fish, upon encountering an unknown species, will pay it cautious exploratory visits. If the stranger never attacks, the visitors become habituated to it and treat it as nonpredatory. However, if the stranger lunges at them, the visitors quickly withdraw and learn, without too much harm to themselves, the identity of a new predator.

In one experiment, naive paradise fish became jaded with the nearby presence of a predatory pike because the pike was kept fully satiated by other means and completely ignored the paradise fish. So the paradise fish came to treat it in the same way a ground-nesting bird might view a cow—as a big placid animal that does not present too much danger. (I'm sure these paradise fish would have experienced a rude awakening had the experiment lasted longer and the pike been allowed to become hungry again.) In contrast, paradise fish that were exposed right from the start to a hungry and active pike quickly learned to avoid it.[25]

Fishes can also acquire predator recognition through social learning. Fathead minnows, for example, are one of those fishes that cannot recognize the predatory nature of a pike, be it by sight or by smell, unless it has had previous contact with a member of that species. Now, imagine that experienced minnows are placed together with naive individuals that have never encountered a pike, and all of these fish are exposed to a flow of water from a tank that holds a pike. What happens is this: the experienced minnows detect the pike smell and react with fear, dashing and seeking cover. Upon seeing them do so, the naive ones also react with fear. The interesting thing is that if the experienced minnows are now removed, and after a few days the pike smell is presented again to the previously naive fish alone, they show a fright reaction on their own. They have learned that the pike odor signals danger, not because a pike attacked them—none did—but simply because they saw other fish show alarm to it.[26]

Such cultural transmission of predator recognition can even take place between species. For example, brook sticklebacks can learn the identity of a predator by watching the fright reaction of experienced fathead minnows. The experiment works also when the smell of a habitat is used instead of the smell of a predator. Fish can learn to recognize the odor of dangerous sites when they are simultaneously exposed to it and to other fish that suddenly show a fright reaction.[27] Earlier we saw that the same result can be obtained with alarm substance rather than the sight of alarmed neighbors.

It goes without saying that knowing the identity of predatory species helps in evading them, if only because the fish know what to expect and can react sooner to an attack. Experiments with zebra danios, minnows, and coho salmon have shown that individuals exposed to simulated attacks by another predatory species early in life survive longer than do unexposed individuals when they are later put in the presence of the real thing, or they act more cautiously when they later see a predator model.[28]

Teaching Hatchery-Raised Salmon to Avoid Dangerous Areas in the Wild

If a fish that possesses an alarm substance system (see chapter 1 on olfaction) is simultaneously exposed to a solution of alarm substance and to the smell of a habitat or fish species it has never experienced before, the fish will associate the new smell with danger and later avoid areas where it can smell that habitat or fish odor. An interesting application of this ability is in the training of hatchery-reared salmonids to recognize their natural predators before release into the wild. Mortality rates are normally very high in salmon and trout that have just been stocked within a stream, and this could be caused by a failure to recognize predators on the part of these naive hatchery-reared fish. But there might be an easy solution to this. Before being sent on their merry way from the hatchery, fish could be exposed simultaneously to their alarm substance and to the water from a tank containing a captive specimen of whatever predator is normally present in the environment. The fish would thus be conditioned to recognize the predator's smell, and hopefully they would be able to remember that smell as indicative of danger and later avoid natural areas in which this scent lingers. Tests performed with juvenile rainbow trout have shown that these salmonids could remember and negatively react to the smell of a pike for at least 21 days after the original conditioning.[a]

As to learning how to evade a predator, there is one interesting experiment on guppies that is worth reporting. In the laboratory of Robin Liley at the University of British Columbia, Wayne Goodey reared guppies under two different conditions: one group was in contact with adult guppies that often chased them, whereas the other group could see or smell adults but could not be chased by them. Later in life, when put in the presence of predatory cichlids, the first group—experienced in being chased, if only by their own conspecifics—escaped from the cichlids more often than did the others. It seems that the experience gained in fleeing from conspecifics could be transferred to another context and used in predator evasion. Even a fry's experience of being chased by a parent and taken into its mouth for safe carrying back to the nest may help the fry to avoid predators later in life, as reported for some populations of sticklebacks.[29]

Finally, learning can also play a role in the recognition of competitors. At the University of Hawaii, George Losey conducted an interesting experiment with the damselfish *Stegastes fasciolatus.* In an outdoor pool supplied with seawater, he let damselfish establish territories. In these territories, damselfish grazed on algae. Meanwhile, Losey also trained Mozambique tilapias to feed either on algae growing on bricks or on freeze-dried zooplankton delivered in midwater by an automatic dispenser. The tilapias were not trained on both food sources simultaneously, and so they became specialists on the one food type they had been trained with. At the beginning of the experiment, Losey placed the algae-covered bricks or the zooplankton dispensers close to the boundary of a damselfish territory, and he monitored the reaction of the damsel to the tilapias that fed there.

At first the damsel showed little aggression toward the tilapias, no matter what they fed on. This is understandable because although tilapias are unknown to damselfish in their natural habitat, they possess body characteristics that are fairly typical of predators and so the damsel gave them a wide berth. Two weeks later, however, the picture had changed somewhat. The damselfish still showed little aggression toward the zooplankton-eating tilapias, but it attacked the algae-eating ones. Through tentative interactions with them, the damsel had learned that those individuals were not piscivorous but were grazers instead and as such were direct food competitors.[30]

This survey of learning in fishes makes it obvious that almost every aspect of a fish's life is susceptible to improvement through experience. Most behaviors in fishes are built on a skeleton of innate (hardwired, instinctive, genetically preprogrammed) instructions. In some cases, this genetic foundation is enough to constitute a full-bodied behavior in itself. In most instances, however, the behavior has to be fleshed out and subjected to environmental fine-tuning. This is where learning comes into play. One of the earliest disputes between comparative psychologists and ethologists was about the relative importance of learning and instinct, with psychologists arguing that behaviors had to be entirely learned and ethologists emphasizing the mostly instinctive nature of an animal's actions. Symbolic of the reconciliation between the two factions, most students of animal behavior now agree that both instinct and learning play interlacing roles in behavior, in the same way that both mortar and bricks (algae-covered or not) are necessary to build a functional wall.

6

Telling Time

Sometimes, when I see people who walk around with no watch on their person, I wonder how they can survive (well, function in modern society, at any rate). I for one need my watch to keep up with a busy schedule of meetings, meals, and social obligations. Now, those carefree people might say that they are leading a more natural lifestyle. After all, in the whole of the animal kingdom, humans are the only ones to wear watches. Ha, I say—not true! In a sense, animals have clocks just like us. They simply do not wear them on their wrists; instead they carry them inside their heads. They have an internal clock that allows them to figure out approximately what time of the day it is, even without looking at the sun or the moon. Even fishes are endowed with this timing mechanism, as illustrated by my favorite species, the convict cichlid.

Cichlids are among the most popular of all pet fishes. To the uninitiated, this may seem surprising because cichlids are pugnacious little beasts, and as such they are not very endearing. But they redeem themselves in that they channel their aggression into a noble pursuit, the defense of their young. For cichlids are among a minority of fishes that provide extensive parental care to their offspring. A male and female cichlid surrounded by a swarm of young fry, constantly ready to attack whatever creature is foolish enough to approach the brood (including fingers from curious people), offer a scene that is, despite all of its implied ferocity, simply charming.

Of all the cichlids, convicts are my favorite because they are fairly easy to obtain from pet shops and to breed in captivity, and they are not shy about performing their parental duties in front of attentive people sitting in front of their aquarium. One behavior so displayed is fry retrieving. When night approaches, parental convicts excavate pits in sand or gravel, and then they seek their free-swimming fry, catch two or three of them in their mouth, return to the pit, and unceremoniously spit the fry into it. The parents do this repeatedly until all fry have been retrieved. By the time darkness falls, all the fry are in one place that is easy to guard and defend against nocturnal predators. Such fry retrieving can be done in the dark at the onset of night (with my infrared goggles, I have verified that it can indeed happen), but obviously it is better to do it in anticipation of night so that full protection is given to the fry as soon as darkness is complete. This interesting behavior begs the question: How do the parents know when the time has come to retrieve? Perhaps they rely on an internal clock, but then again maybe they simply use the advent of dim crepuscular light as a cue to the approach of night.

To answer this question, I have kept convict parents in windowless rooms in which programmable timers turned off all the lights at once every evening. No crepuscular signal for them. Interestingly, many of the parents still started to retrieve their young ahead of time, starting 15–20 minutes before the lights would go off, even though there was no external sign that nightfall was impending. This behavior supports the notion that an internal clock is at work in parental convicts.

Admittedly, in another room that lacked windows but in which I provided a dusk signal by turning some of the lights off every day at 15 minutes before nightfall, fry retrieving took place more assiduously.

Light cues can therefore modulate the intensity of fry retrieving. Even in that room, however, I found evidence of an internal clock. When I turned off some of the lights at different times of day, the parents would not retrieve during midday or midafternoon "dusks." They would do it only during the dim pulse that immediately preceded night. It seems an internal clock "told" the convicts that the midday or midafternoon episodes of dim light were not at the right time of day to be the true precursor of night; instead, the dim light might correspond to a dark storm or to a bout of foraging underneath a riverbank. Another way to look at this mechanism is to pretend that the internal clock opens a "gate" only at dusk, a window of time during which retrieving can be performed; within that window, the levels of retrieving are influenced by light intensity, and they are higher when the light is dim.[1]

Parental cichlids are not the only fishes that need to anticipate the arrival of night. In both marine and freshwater ecosystems, entire guilds of diurnal fishes get on the move at dusk to reach their distant sleeping quarters and prepare to spend the night there. Nocturnal guilds do the same at dawn.[2] Such twilight migrations are launched in advance of full light or full darkness and may therefore be controlled by an internal clock. Experimental manipulations of the lighting regime such as the one described above would be necessary to test this idea. Unfortunately, it is unlikely to be done any time soon, given the near impossibility of building aquarium facilities large enough for the expression of daily migratory behavior.

One manipulation that is easier to perform in the laboratory is the delivery of food at specific times of day. Automatic feeders can easily be programmed for this task. Scheduled food delivery may have natural counterparts in the real world if we are willing to imagine that some fishes feed on insects that are active only at specific hours of the day, or on phytoplankton that is more nutritious at specific daily times.[3] At any rate, there is ample evidence that fish can anticipate the arrival of food when it is delivered at the same time every day (see fig. 6.1).[4] The first serious study on this topic took place in the 1960s. At the University of Michigan, Roger Davis and John Bardach fed Atlantic tomcod, scup, and mummichog at the same time every day and found that the fishes became noticeably more active 2–6 hours before mealtime but not at other times. (Activity, by the way, was recorded by connecting an electrical switch to a network of rubber bands stretched across the aquarium, which the fish kept hitting while swimming.)[5]

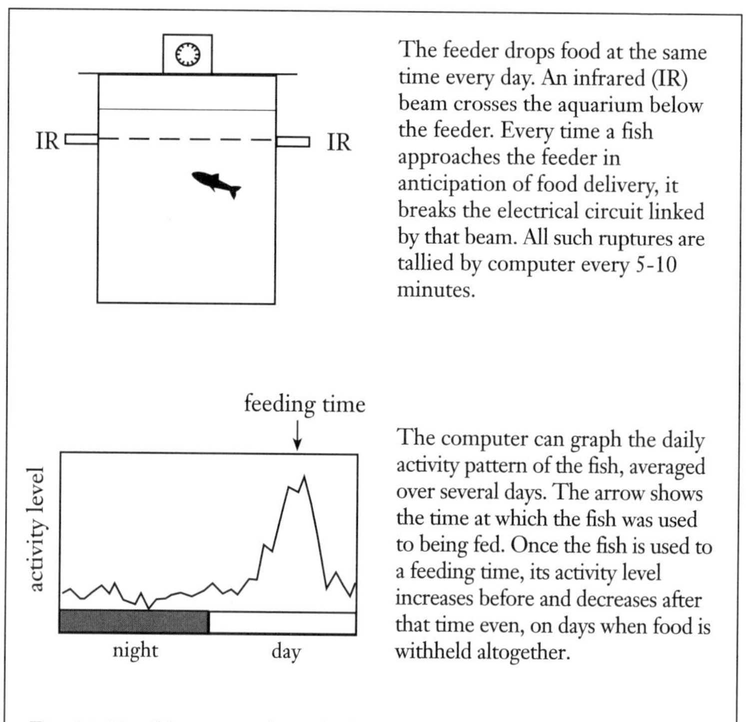

Fig. 6.1. Possible setup and results for an experiment on food anticipation in fish.

Since then, many more species have been tested for food-anticipatory activity. Chief among them is the goldfish, extensively studied by Richard Spieler during his tenure at the Milwaukee Public Museum. In one experiment, Spieler and colleague Teresa Noeske fed some goldfish only at night, others only during the day, and yet others near the time of lights-on or lights-off (the photoperiod was artificial; the fish could not see the sun). Then, after a few weeks, the food was withheld for 10 days. During that period of fasting, each fish turned out to be active only around the hour at which they used to be fed, be it during the day, the night, or near lights-on or lights-off. Activity started to rise a few hours before normal feeding time. This could not be a direct response to food because food was no longer delivered. Instead, it seems that the fish had learned to adopt a daily phasing of activity that maximized their foraging success. Diurnal, nocturnal, crepuscular, they did not care, as long as they could normally get food.[6] (I know of a similar manipulation of activity phasing that has been attempted with a nocturnal mammal, the weasel. Food was made much easier to obtain during the day, but the weasel remained stubbornly nocturnal, showing less plasticity than goldfish.)[7]

In Spieler's phasing experiments, it is remarkable that goldfish decreased their level of activity past the normal time of feeding, even during the fasting days when no food was delivered. If not for this, it would have been possible to interpret increased activity before mealtime as an expression of growing hunger rather than knowledge of time. But the goldfish were not fed on any of the fasting days, yet they still decreased activity once the normal time of feeding was passed. They were still hungry (no food had been delivered), but their clock told them that feeding time was over and that, until the next day, it was useless to expect food and look around for it.

There could be several advantages to food anticipation. First, more movement before mealtime may enable a fish to be the first one to get to the food. Second, the physiological machinery necessary for the digestion and assimilation of food can be warmed up in advance, so that it is in an optimal state as soon as the food arrives. And third, the fish can avoid foraging at times that are unlikely to provide rewards. However, I am unaware of any study that has tried to prove these ideas with fishes.

Fishes can do better than anticipate just one daily mealtime. They can learn at least two different mealtimes and can even associate each one with a different location. I have been able to demonstrate this with my second favorite species (after the convict cichlid)—the golden shiner, a shoaling minnow that in nature roams over vast areas in lakes. Golden shiners are easy to keep in the laboratory, and they feed readily on commercial food flakes delivered at the surface by automatic feeders. They also seem to be very smart when it comes to learning about feeding time, even under an artificial light–dark cycle with no view of the sun.

In a windowless room, I placed groups of shiners in long aquaria divided into two sections by incomplete partitions that blocked most of the view but still allowed passage from one side to the other. For three weeks, feeders dropped food on the left side of the aquaria in the morning and on the right side in the afternoon. I reckoned that three weeks would be long enough for the fish to learn the association between daily time and place. Then, on a test day, no food was given, and with a camcorder set on time-lapse I recorded the position of the fish throughout the whole day. As I had hoped, most fish in the group took positions on the correct side at the correct time—left in the morning and right in the afternoon—even though no food cue was available. The experiment worked again when I trained the fish to be on the left side in the morning, on the right at midday, and back on the left in late afternoon. This mirrored the movement pattern of the fish in their native lake, as they are known to move from open water to the littoral zone after dawn, and back to open water in late afternoon.[8]

This capacity for time–place learning hints at a certain level of sophistication on the part of the internal clock, namely, that it can be consulted more than once during the day. In straight night or day anticipation, or single mealtime anticipation, the internal timing mechanism of the fish needs only to work like an alarm clock. It needs to give a signal only once a day. The arrival of day, night, or food is a single daily event. But with time–place learning, we see that the clock can be consulted at least twice a day.

In fact, ethologists know that the clock of fishes can be consulted more or less continuously throughout the day. This is because they know that fishes possess a sun-compass mechanism that allows them to find their way back home when they get lost.

Most people are familiar with the notion that with a watch and a view of the sun, it is possible to infer the position of any cardinal point. The sun is always over the east in the morning, over the south at midday (or north in the Southern Hemisphere), and over the west at the end of the day. Points in between can be interpolated. For this mechanism to work, it is absolutely essential that a continuously consulted clock be available. So, if we can demonstrate that fishes are capable of sun-compass orientation, we will have shown that they have a sophisticated internal clock.

A classical demonstration of sun-compass orientation goes like this: first, a fish is placed inside a container in the middle of a circular pool. The surroundings are uniform except for a view of the sun or at the very least a bright lamp that moves around the room like the sun. The pool itself is as uniform as possible, and it is regularly rotated to prevent the fish from relying on small landmarks inside the pool that might not be perceivable by people. All around the periphery of the pool are a number of identical shelters or feeding stations, depending on the motivation of the fish at the moment of the test (finding shelter or finding food). They too can be rotated regularly. It is assumed that the fish always wants to go in the same cardinal direction, either because it does so in nature (some fish, for example, go offshore to feed and inshore to take shelter, and these represent constant directions throughout the life of a fish whose home range remains the same) or because the individual has been previously trained by the researcher to always go in the same direction (for example, of all the shelters around the periphery, only the one to the southeast is open).

So, the fish is released from the central container, and the direction in which it swims is noted. If the fish can use sun-compass orientation, it should always swim in the correct direction, irrespective of the time of day at which the test is conducted, provided that the sun (or lamp) has moved around the pool at a natural rate. Machiavellian researchers can also hide the true position of the sun and use mirrors to deflect its apparent position by, say, 90°, with the expectation that the preferred direction of the fish would also shift by 90°. Another variant is to test the fish with and without a view of the sun (on sunny and cloudy days, for example), with the expectation that correct orientation would be lost when the sun is not visible.

With these methods, sun-compass orientation has been demonstrated in at least a dozen species of fish: white bass, pumpkinseed sunfish, bluegill sunfish, green sunfish, largemouth bass, Southern starhead topminnow, sockeye salmon, mosquitofish, two cichlids (*Cichlaurus severus* and the uaru), and two parrotfishes (the purple and the rainbow parrotfish).[9]

In at least one of these studies, the role of the internal clock was elegantly confirmed. Phillip Goodyear and David Bennett, then at the Savannah River Ecology Laboratory in South Carolina, captured immature bluegill sunfish that were in the habit of moving in a known direction to reach their natural refuge. When tested in a circular pool at midday in full view of the sun, these fish oriented correctly, equating the sun position with a southerly direction and moving appropriately relative to that. But Goodyear and Bennett also kept some fish in the laboratory under a photoperiod that had been advanced by 6 hours (the fish "got up" at 1:00 A.M. instead of 7:00 A.M.). When these fish were tested under the natural midday sun, they did not orient as if the sun was over the south. Instead they interpreted the position of the sun as west, even though the sun was high in the sky. Their internal clock, which had been advanced along with the artificial photoperiod, told them they had been up for 11 hours and that this was the end of the day, and every decent fish knows that the sun is over the west at the end of the day, no matter how high and bright it is. These results clearly indicate that fishes do not use sun height as a temporal cue, but sun position and the counsel of their continuously consulted internal clock instead.[10]

Orientation in Fishes

In this book so far, we have touched on elements of another good cognitive ability of fishes: the capacity to orient properly in a vast expanse of water to find home. To help find their way around, fishes can memorize the scent of home, learn the visual landmarks that characterize home, listen for sounds of home, and figure out geographical direction based on magnetic compasses or sun compasses.[a]

The traditional way to prove that fish can find home is to offer them an all-expenses-paid trip to some far away destination (by their standards) and see if they can come back on their own. In New Zealand, mottled triplefins (small fish only 7–10 cm long that spend all of their adult life on the same two square meters of territory) have been displaced more than 700 m (or roughly the length of seven football fields) along a rocky reef, and most of them came back within 4–6 days. Some other feats of home-finding include radiated shannies that traveled more than 270 m to return to their home base, flathead catfish that traveled more than 1 km, various sunfish and bass that traveled more than

3.5 km, yellowtail rockfish that traveled more than 22 km, and one skipjack tuna that traveled more than 35 km to return to its usual resting place.[b]

Some researchers fit their displaced subjects with ultrasonic transmitters so that the return route taken by the fish can be plotted. But these transmitters are expensive, have a lifetime of no more than several weeks, and are so bulky that their use is restricted to large fishes (sharks, for example). For fun, compare this methodology with the 1950s precursor of ultrasonic telemetry, a method that was low tech and still worked well for species living in lakes. Hooks were simply inserted through the dorsal musculature of the fish and attached to a small float via a long thread. The movement of the float at the surface would betray the movement of the fish below and could easily be charted.[c]

Night anticipation, food anticipation, time–place learning, and sun compasses have been documented in all vertebrate classes, not only in fishes. The internal clock involved is likely to be more or less the same in all; it is the circadian clock, the one that runs spontaneously with a periodicity of approximately 24 hours. (*Circadian* comes from the Latin *circa*, or "about," and *dies*, "day.") Most animals, including humans, have a circadian clock. Researchers are not too sure where in the brain this clock is located for fishes, but we know its location in mammals—it is in the hypothalamus. (Touch your palate with your finger and you're almost touching your internal clock.) This is the clock that makes us feel sleepy in the evening or that spontaneously wakes us up in the morning.

Maintaining the temporal integrity of sleep may have been one of the primary functions of the circadian clock in animals. The argument can be expounded thus, using fishes as an example: The eyes of many diurnal fishes are specifically suited to well-lit environments, and consequently these fishes do not operate at their best in the dark. It is better for them to stay inactive at night and save energy, that is, to sleep. But how should sleep be triggered? Simply falling asleep in response to darkness is not a good mechanism because fishes sometimes encounter places that are permanently dark (underneath the ledge of a riverbank, for example); if fishes were to fall asleep automatically then, they would be transformed into sleeping beauties, never to wake up again. For nocturnal species, falling asleep simply in response to light is no good either. If a nocturnal fish were to linger outside of its shelter at dawn and fall asleep automatically just because of the emerging light, it would spend the whole day out in the open, exposed to diurnal predators.

For both diurnal and nocturnal species, it is better to have an internal clock that tells the fish when it is time to become active and inactive. We can therefore conceive of the circadian clock as a timer switch that turns the sleep drive on and off at the right time of day. The result is a fish that displays a long bout of activity alternating with a long bout of sleep or inactivity, with a periodicity that approximates that of the light–dark cycle. The light–dark cycle can synchronize the sloppy periodicity of the circadian clock back on track at 24 hours, but it is important to remember that the light–dark cycle does not generate the sleep–wake cycle. The latter is expressed even in constant lighting conditions—constant light or constant darkness—which shows that the switch between sleep and wake is triggered internally, not in response to an environmental signal.

Some of the best examples of stable activity rhythms in fishes kept under constant conditions have come from the graduate studies of Martin Kavaliers at the University of Alberta.[11] Kavaliers used an ultrasound system to monitor locomotion in various freshwater fishes kept in aquaria. He placed diurnal lake chubs in constant darkness and found that their activity patterns remained rhythmic, one block of activity alternating with one block of inactivity every 25–28 hours (see fig. 6.2). He obtained similar results with mummichogs and white suckers, also showing that fish in groups displayed a better demarcation between activity and inactivity, and a more regular rhythm, than did single individuals.

These self-sustained rhythms did not last for very long, however. After a maximum of approximately 20 days, the chubs and mummichogs started showing irregular bouts of activity throughout the 24-hour period. Some nocturnal species seemed to do better in that respect: burbot could remain nicely rhythmic for up to 40 days before breakdown, and the same has been noted in other laboratories for

Fig. 6.2. Actograms are depictions of daily activity in animals. Each line represents a day from midnight to midnight, and successive days are plotted one below the other. Dark rises on the line are commensurate with a fish's (or group of fishes') level of activity as recorded by an automated system. If the animals are kept in constant conditions, such as constant darkness for 24 hours a day, they "free-run," that is, their internal clock drifts and the animals get up later and later every day. The fictional example illustrated here would be rather nice for fishes, which seldom yield such clean and long-lasting rhythms. The main exceptions are nocturnal species; their rhythms seem to be more robust.

hagfishes, catfishes, lamprey, and electric fishes.[12] Nevertheless, we are still far from the year-long records of stable rhythms that have been extracted from birds and mammals, so we must therefore conclude that left on their own, away from the light–dark cycle, the internal clocks of fishes can maintain the integrity of the sleep period for only a limited time.

A different kind of support for the view that circadian clocks and sleep integrity are functionally related comes from a comparative study of sharks conducted by Donald Nelson and Richard Johnson off the coast of California. These researchers observed two species of nocturnal bottom-dwelling sharks, the California horn shark and the swell shark. Nelson and Johnson found that the horn shark was still fairly alert by day—not a sound sleeper—but that the swell shark was very lethargic—a good sleeper. One individual of each species was brought into the laboratory to measure activity patterns, and interestingly enough, the nonsleeper was found to have no discernable circadian rhythm of locomotor activity under constant conditions, whereas the good sleeper showed a strong one.[13] We may surmise that animals that do not sleep do not need a daily timer; therefore they did not evolve a circadian clock, and they cannot show spontaneous daily rhythms of activity.

Along the same lines, it is interesting to note that obligatory cave-dwelling fishes, which live in a place in which darkness always prevails, do not show circadian activity rhythms either.[14] Because their world does not operate on a 24-hour basis, these fishes have no need for a circadian clock. If we subscribe to an ecological view of why animals need to sleep (to save energy and keep a low profile during the half-day for which they are ill adapted), then we can see that cave-dwelling fishes need not sleep at any particular time—indeed, they may not need to sleep at all—and that therefore they do not require a timing mechanism for that activity.

Do Fishes Sleep?

University students are sometimes paid to spend a few nights in sleep research laboratories, where their sleep patterns are monitored. Researchers recognize the onset of sleep because the eyes of the subject are closed and distinctive patterns in the electrical activity of the brain are picked up from the surface of the skull.

Fishes are never paid to spend time in sleep labs. Their eyes are always open (they have no eyelids), and they lack the complex brain structures that generate the electrical pattern diagnostic of sleep in mammals. So how can one measure sleep in fishes? In fact, do fishes sleep at all? And who would care to know anyway?

Curious fish lovers might care enough to try and measure sleep by using behavioral criteria. With infrared equipment they could—as I have done—observe that cichlids are sluggish and unresponsive at night, resting on the bottom with their eyes cast downward, with a lower respiratory rate and little reaction to the introduction of food. Such behavior is certainly suggestive of sleep. Following in the footsteps of a team of Russian researchers, curious aquarists could poke catfishes with glass rods at night and realize that these fish do not seem to notice.[d] Like many divers, they could observe fishes at dusk "settling down for the night," snuggling into the same hole, crevice, sponge, weed patch, or coral head night after night. Such places are to fishes what beds are for us.

Not all fishes "sleep," however, and those that do may not do so in a nice and orderly manner like us. After a few long days and nights of sustained observation, researchers in South Africa came to the conclusion that their juvenile tilapia showed no sign of sleep (although the adult fish did).[e] With my infrared goggles, I have seen that my convict cichlids, although sluggish at night outside of the reproductive season, can continue fanning their eggs and wrigglers 24 hours a day (human parents of newborn babies no doubt would be envious).[f] I have also set up minnow traps in a small stream near my laboratory and caught lake chubs only during the day in midsummer, which suggests that they might sleep at night, but to my surprise the capture pattern became exclusively nocturnal at the time of in-stream or out-stream migration. Perhaps during migration the chubs have to pass over shallow riffles, where diurnal fish-eating birds might easily pick them up if they did not travel under the cover of darkness. Therefore, it is better to eschew nocturnal sleep and move at night.[g] Flexibility in sleep patterns is also illustrated by salmon, which are known to be diurnal when the temperature is warm but nocturnal when it is cold. Perhaps they catch more food under the better lighting conditions of the day, but only when the water is warm and their reflexes are good are they willing to face the diurnal warm-blooded birds that would like to make a meal out of them.[h]

Do fishes sleep? I think most of them do, although I would qualify the statement by adding that fish sleep is not exactly the same sort of state that we humans are used to experiencing.

Aside from the circadian clock, there is another timing mechanism that some fishes may have. It is one that they share only with those few animals that live along shorelines. This clock runs at a periodicity of approximately 12.4 rather than 24 hours, which corresponds of course to the duration of the tide cycle. Such a clock is said to be circatidal, and it allows an animal to anticipate high or low tide events. Many species, including gobies, blennies, flatfishes, and mudskippers, have been plucked from their intertidal habitats and transferred to constant conditions in the laboratory; there they have exhibited a circatidal rhythm of activity, moving around for a few hours every 12.4 hours or so, in fairly good synchrony with the tides outside. But as with circadian clocks, circatidal rhythms in animals do not persist for long under constant conditions. The rhythms disappear after a few days. However, if after the rhythms have vanished the animal is exposed to only two simulated tidal cycles with their attendant changes in hydrostatic pressure and wave action, this is enough to get the rhythms going again, if only for a short time.[15]

So, time and tide wait for no one, not even fishes. Internal clocks have evolved in animals because of the intrinsically rhythmic nature of the physical world on Earth, and although it is true that environmental cycles tend to be more attenuated in water—temperature does not vary as much as it does in air, for example, and daylight does not appear as bright—such cycles are still present and important enough to create a necessity for fish clocks to keep on ticking.

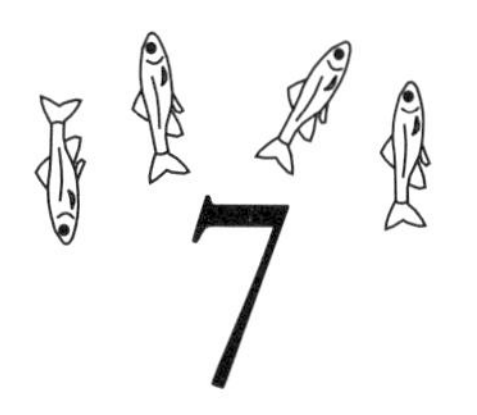

7

Individual Recognition

In general, it is not too hard to tell fish species apart. Experts are able to distinguish most fish species solely on the basis of physical characteristics. However, the task of recognizing individuals within the same species is not so straightforward. Human observers will often deem it impossible to distinguish between, say, two cardinal tetras of similar size or two adult minnows. This is why so many methods for tagging fish have been developed by behavioral and fisheries scientists.

The fishes themselves, however, have a much easier time at it. There is little doubt that many of them have the cognitive capacity to differentiate and remember individuals. To convince oneself of this, one need only consider species that form pair bonds. In some fishes (for example, cichlids and anemonefishes), males and females get together during the reproductive season, court each other, and eventually form breeding pairs. The pair bond between them is often strong and long-lived. Together, the male and female defend a territory, mate, and take care of the progeny. While tolerating each other, they actively chase away other members of their own species. Obviously, all of this could not happen if the two members of the pair were unable to recognize each other.

Hans Fricke is a well-known ethologist who has conducted many behavioral studies on marine fishes in tropical seas. In one of these studies, he built an enclosure with four arms radiating from it. Removable partitions isolated the arms from the central section. This apparatus was lowered to the bottom of the Gulf of Aqaba in the Red Sea. In the central section, Fricke inserted an anemone and a resident anemonefish, *Amphiprion bicinctus*. In the arms, he placed other anemonefish, one of which was the regular mate of the central resident. He then lifted all the partitions. Upon seeing the central anemone, all of the peripheral fish tried to reach it. The resident fish vigorously defended its anemone, attacking all newcomers with one exception: not a single attack was leveled at its mate. Fricke then repeated the experiment, but this time he masked visual cues, either by uniformly dying the peripheral fish with the colorant Bromo-cresolgreen or by enclosing the fish in green containers. Under these new conditions, all fish were attacked, including the resident's mate. Fricke concluded that visual cues are important for mate recognition in anemonefish, although it must be said that his masking techniques may also have screened any odor given off by the fish. There was no evidence, however, that the central fish took the time to sniff at the intruders before attacking them.[1]

Olfactory cues may nevertheless be important for mate recognition in other species and in other contexts. Take the case of convict cichlid mates interacting at night. In the laboratory, I gave a pair of convicts access to a 20 × 10 × 10-cm nest box that could only be entered through a hole in one of its walls (another one of its walls was solid but transparent so that I could see inside). When the young were at the free-swimming fry stage, I started a night experiment. Just before the time of lights-off in the evening, I removed the male from the aquarium. This did not prevent the remaining female from retrieving her young and watching over them inside the box at night. There are nocturnal predators in the convicts' natural environment, and parents have been naturally selected to keep watch all night long. The point here is that other convict cichlids are sometimes keen on devouring young fry that do not belong to them. The female should therefore expel from her nest all convicts but her own mate, provided she can tell the difference. The question is: Can she, given that it is completely dark?

After 3 hours, in the complete darkness of night, I put on infrared goggles and introduced various males one at a time into the nest box. The male could be the female's mate, or it could be a strange male that was either smaller, the same size, or bigger than the female's mate. Under infrared light, I saw that the female never attacked her mate, but she directed numerous headshakes, pushes, and bites to the other males, irrespective of their size. So she seemed capable of mate recognition. I also noticed that she started her attacks only after her snout came into contact with the intruders. This suggested to me that she used short-range chemical cues for mate recognition; obviously she could not use visual cues because the experiment took place at night.[2]

Of course, nothing prevents a cichlid from also using visual cues if conditions allow it. During the day, female convict cichlids often charge at territorial intruders after detecting them from such a great distance that it is doubtful any chemical cues could have had time to travel between the two. The same females leave their mate in peace at a similar distance. They can therefore recognize their mate visually, and only at night are they forced to rely on olfaction.

Recognition can extend to more than one breeding partner. In the cichlid *Lamprologus brichardi*, young fish do not always leave home after reaching independence. Some remain behind to help their parents defend the natal territory. They will forego the opportunity to breed on their own, electing instead to help their parents rear a new brood. The parents must be able to tell these helpers apart from other juvenile cichlids, whose intentions toward the new brood are anything but nurturing (as in convict cichlids, juvenile *L. brichardi* are fierce egg and fry predators).

Helper recognition has in fact been demonstrated by Eva Hert at the Max-Planck-Institut für Verhaltensphysiologie in Germany. An aquarium housing a pair of parental cichlids was placed side by side with two smaller tanks, one tank containing one of the parents' helpers and the other containing a helper from another nest. To make sure that the first helper did not react to the sight of its parents, thereby giving behavioral cues as to its identity, one-way mirrors were inserted between the tanks. The parents could therefore view the helpers, but the helpers could not see the parents (see fig. 7.1). Hert reported that although the parents displayed aggressively toward both fish (who themselves were somewhat aggressive because they could see their own reflection in the mirrors), the displays were less numerous in front of the legitimate helper, implying that the parents had recognized their helper individually.[3]

Parents can recognize each other, but can cichlid fry recognize their own parents? The answer is no, but not because they have no discriminative powers, as we shall see. It is just that these powers do not extend to recognition of individual parents. If an aquarium with cichlid fry is sandwiched between one that contains their own mother and one that contains the mother of a different brood, the fry simply elect to stay close to the most active adult, and this is not always their own mother.[4] Similarly, if cichlid fry are exposed to two water inflows that have different odors, they will consistently prefer the smell of their mother over blank water, the smell of their father over blank water, and the smell of their mother over that of their father, showing that they do have discriminative powers based on olfaction, but they will not prefer the smell of their own parents over that of other parental adults.[5]

The absence of parental recognition in cichlids can be illustrated even more dramatically: in at least one study, convict cichlid fry were

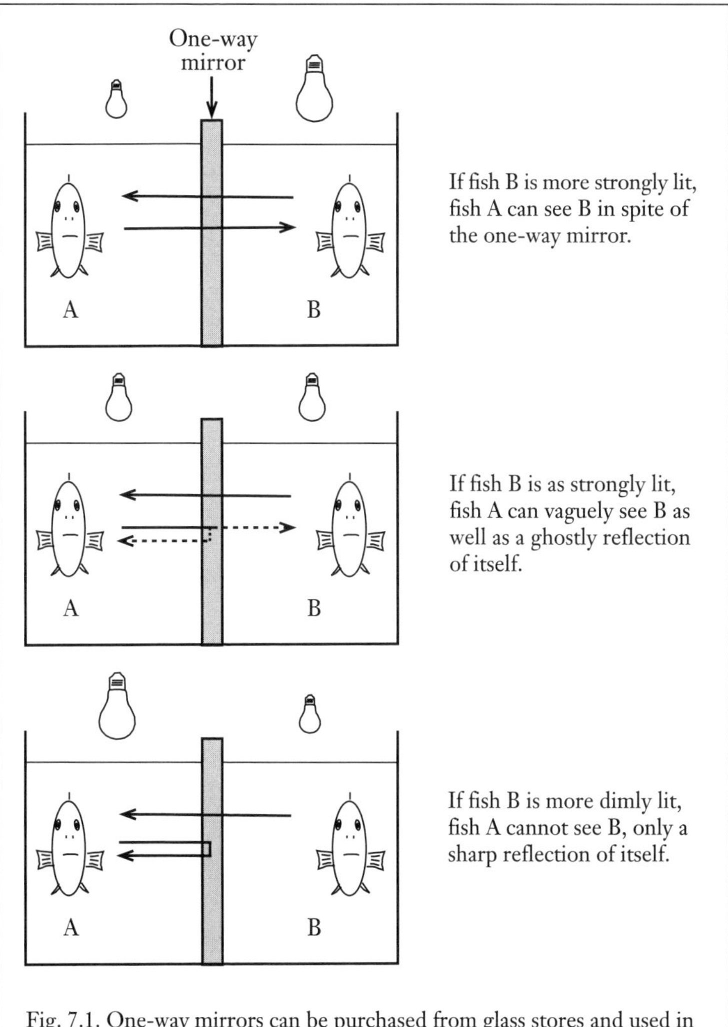

Fig. 7.1. One-way mirrors can be purchased from glass stores and used in some experimental protocols. From one direction, they are transparent (above, fish B can always see fish A). From the other direction, they are either transparent or reflective, or a combination of the two, depending on the relative light intensity on either side, so that fish A may or may not be able to see fish B.

easily persuaded to follow mechanically moved dummies that were only crude replicas of their own parents. The young were not complete fools, for they preferred dummies that bore the markings of their own species to those of other cichlids. Yet their willingness to follow a robot does not hint at filial loyalty, let alone individual recognition.[6] Mind you, being unable to recognize their parents may not matter to fry, given that many cichlid parents are known to accept foreign fry under the umbrella of their protection, as long as there is not too great of a size difference between foreign and legitimate young. For fry separated from the rest of their kin, following an attack by a predator, for example, "any port in a storm" might well represent the best behavioral strategy.

Aside from marital bliss, another aspect of fish behavior made possible by individual recognition is the establishment of stable dominance hierarchies and territorial boundaries. In hierarchical groups, individuals would be forever fighting among themselves if they could not recognize their rivals and remember the rank each one occupies within the pecking order. And in the case of territorial species, recognition of neighbors means that fish do not have to forcefully reassert property rights every time they meet at their territorial boundaries.

Some of the first demonstrations of individual recognition in fishes came from experiments with aggressive species (trout, swordtails, and paradise fish, for example).[7] The basic procedure was to pit two fish against each other and let them fight it out to determine their respective dominance status. Then the subordinate fish was removed and the dominant was left alone. Some days later, this dominant fish was either reintroduced to the individual it had defeated earlier or introduced to a newcomer that had also been defeated earlier but by another fish. The usual outcome was that less aggression took place between two protagonists that had already met compared with the aggression between two that had never seen each other before. This repressed antagonism in front of familiar rivals is another form of the "dear enemy" effect I presented—also in the context of individual recognition but this time based on acoustical or electric signatures—in previous chapters. The existence of a dear enemy effect attests to the fact that familiar adversaries can recognize each other.

Out of Sight, Out of Mind

Memory of recognizable rivals is but one way to abate aggression between territorial neighbors. Another approach corresponds to the expression "out of sight, out of mind." The less often neighbors see each other, the less tempted they are to pick a fight. Several studies have shown that in habitats that are structurally complex, with lots of vertical relief, plants, or plates installed by curious experimenters, more territorial fish can cohabitate next to one another, and there are fewer belligerent interactions among them.[a]

The mudskipper *Boleophthalmus boddarti* is energetic in its approach to cohabitation. This fish occupies permanent territories on the intertidal mudflats of Kuwait. At high fish density, and therefore under high intruder pressure, mudskippers start to build mud walls 3–4 cm high at the boundaries of their individual territories. They do so by carrying mouthfuls of mud that they deposit along boundary lines. These walls prevent other mudskippers from crawling into the territory. Through the simple device of shielding the neighbors from view, walls also save the territory owner the trouble of worrying about rivals and bickering with them. In one experiment, wrecking crews (naughty researchers) destroyed the walls on one side of a mudskipper's territory. The fish soon started to build the fortifications anew, but it often paused to display and fight with its now-visible neighbor. Peace returned only when the wall was completed.[b]

By the way, the above is one of the few examples of construction in the fish world. Others would be the nests of male sticklebacks (plant material stuck together with glue produced by a special gland) and gouramis (air bubbles blown by the fish and floating together at the surface) or the bowers of some African cichlids (sand mounds built by males and used by females to evaluate the quality of potential mating partners).

Yet another example comes from mudskippers again. The species *Periophthalmodon schlosseri* lives in burrows dug in the mud of intertidal zones in Southeast Asia. This fish excavates special chambers within its burrow. A mudskipper can then take air into its mouth, bring it down below, and store it inside the chambers. Oxygen can diffuse from this pocket of air into the water of the burrow, or it can be tapped into directly if for whatever reason the fish (or its young, which are born in the burrow) deem it too dangerous to emerge at the burrow entrance.[c]

Recognition of familiar enemies can be based on visual cues alone. At Queen's University in Ontario, Joe Waas investigated the behavior of territorial three-spined sticklebacks toward their neighbors. Waas had three aquaria positioned end to end, each one with a male stickleback in it. The middle fish was a "resident" that was regularly exposed to females to incite him to build a nest and become territorial. This fish could see its two neighbors, but these in turn could neither see him nor the females with him because the aquaria were separated by one-way mirrors. After 4–6 days of this restricted visual contact, Waas conducted a test in which he either switched the position of the neighbors (that is, the left neighbor went to the right and the right neighbor went to the left) or he performed a "mock" switch (the neighbors were dip-netted but put back into their original tanks). He then sat down and observed the reaction of the resident fish in the middle. That fish charged and bit at the side glass significantly more often after a true switch than after a mock one. It seemed that the resident had perceived the change in neighbors, had interpreted this as a territorial shuffle, and had moved to reassert his boundaries. The middle fish's perception of the change implies that the neighbors were recognized individually.[8]

Experiments of this kind can even be conducted in the field. For example, one can select a territorial individual and then capture one of its neighbors as well as one stranger from farther away. These two captured fish are put into clear bottles, which are then presented to the resident fish from various sides of its territory. If the regular neighbor is presented from the side at which its own territory lays, little aggression by the resident ensues. But if the neighbor is presented from an unusual side, or if the stranger is presented from the side at which the regular neighbor normally lives, then all hell breaks loose. Such an experiment has been performed with success by using the three-spot damselfish. As with the results with sticklebacks, these results can only be explained in the light of individual recognition of territorial rivals and memory of their usual location.[9]

In some species, rival recognition can operate through olfactory channels only. John Todd, whose work on catfish I have already alluded to in chapter 1 olfaction, reported the following anecdote. Two small yellow bullheads lived next door from each other, or in other words, within contiguous territories inside a large aquarium. They both excluded each other from their respective territories. However, both had had the previous misfortune of being attacked and strongly dominated by the same larger individual. When water from the aquarium in which this larger catfish now lived was poured into that of the two smaller fish, these two suddenly seemed to be gripped by fear. They fled and, forgetting their differences, hid together in the same overturned flowerpot (see fig. 7.2). Using water laced with the scent of another large catfish that the two small ones had never met

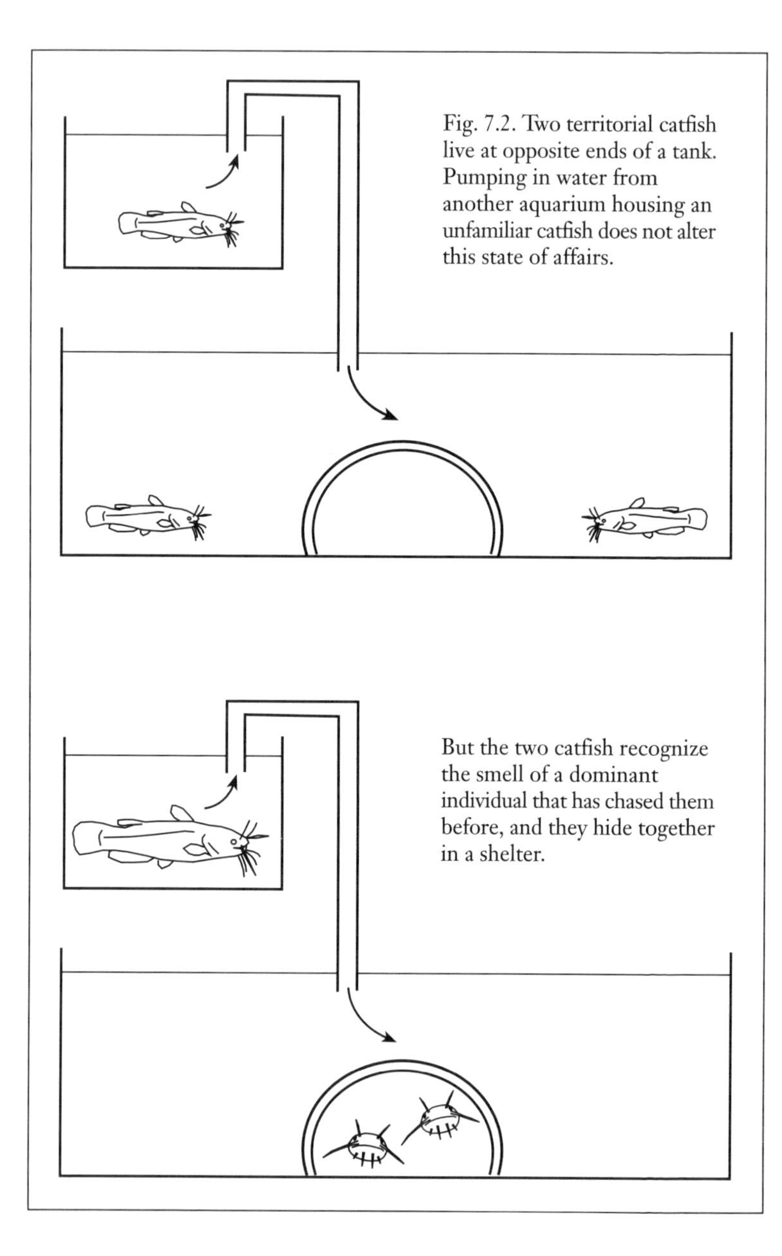

Fig. 7.2. Two territorial catfish live at opposite ends of a tank. Pumping in water from another aquarium housing an unfamiliar catfish does not alter this state of affairs.

But the two catfish recognize the smell of a dominant individual that has chased them before, and they hide together in a shelter.

before did not elicit this fright reaction. Obviously, the small catfish had memorized the smell of the tyrant and could still distinguish it from that of other bullheads.[10]

This experiment is not the only one that John Todd conducted on the olfactory interactions of these catfish. In association with John Bardach and Jelle Atema, he also used a classical training procedure to demonstrate individual scent discrimination in bullheads. He had two "donor" bullheads living in separate aquaria. Water from these aquaria was collected, and 50 ml of it was poured once in a while into the tanks of other "experimental" bullheads. Each pouring was almost immediately followed either by the delivery of food at the surface or by an unpleasant electric shock. (Alert readers will recognize a Pavlovian conditioning protocol here.) For some experimental fish, water from donor A was always followed by food, whereas water from donor B was followed by a shock. The reverse pattern was applied to the other fish.

After about 30 such exposures, all experimental fish successfully learned to associate each donor's odor with its normal consequence. When presented with the "food" catfish odor, the bullheads rose immediately to the surface, where food normally appeared. When they perceived the "shock" catfish odor, they retreated to a shelter that protected them from the shock. They remembered this association for at least 3 weeks, even without retraining. Moreover, the experiments worked almost as well when the signal was simply skin mucus dissolved in water. However, the experiments ceased to work when the subjects were fish that had had their olfactory receptors destroyed. All of this shows that bullheads can recognize different individuals, that they can do so based on their sense of smell only, and that the recognized odor comes in good part from skin mucus.[11]

Lee Dugatkin champions the view that the cognitive abilities of lowly fishes are more elaborate than we might think and that fishes may use these abilities in strategic contexts. His research has provided two examples in which fishes recognize other individuals and choose to associate with those that provide less competition for food (that was in sunfish) or less competition for mates (in guppies).

In the sunfish experiment, Dugatkin and his co-worker David Wilson, from the State University of New York at Binghamton, established a shoal of six bluegills. At regular intervals, the researchers took two fish from the shoal and placed them together in a separate tank in which 20 pieces of mealworm lay scattered on the floor. They noted how many pieces each fish took. They kept doing this until all possible pairwise associations had been tried within the shoal. Then they arranged three aquaria end to end and repeatedly placed one of the fish (the "test" fish) in the middle and one "stimulus" shoalmate in each of the two end aquaria. The question was: Would the middle fish consistently spend more time on the side of the shoalmate with whom it had had the most success in competing for food?

By and large that is what happened. If a test fish had previously obtained, say, 12 pieces of food while foraging with stimulus fish A but only 8 pieces while foraging with fish B, then in the subsequent preference test it spent more time next to A. However, if the preference test involved a regular shoalmate versus a stranger from another shoal, the test fish chose to spend more time next to the regular partner, irrespective of its former success at getting food with that partner. Dugatkin and Wilson concluded that bluegill sunfish can visually recognize individuals, that they prefer to be with recognized shoalmates over unrecognized strangers, and that in a choice of two recognized shoalmates, they strategically pair up with the one that is a poorer competitor for food. Visual cues must be sufficient for this recognition to take place, because all fish were in separate aquaria during the preference tests (see fig. 7.3).[12]

With a different colleague, this time Craig Sargent from the University of Kentucky, Dugatkin wondered if male guppies would recognize and prefer to associate with other males that seem to be less attractive to females, making themselves "look good" by comparison. His experiments involved one female and three males (male 1, male 2, and a test male) and three stages. In the first stage, adjacent aquaria were positioned in such a way that the test male was close to a female and could see male 1 in the distance (the idea was that male 1 would then be perceived as a "loser"). In the second stage, the test male was alone and could see a female next to male 2 in the distance (male 2

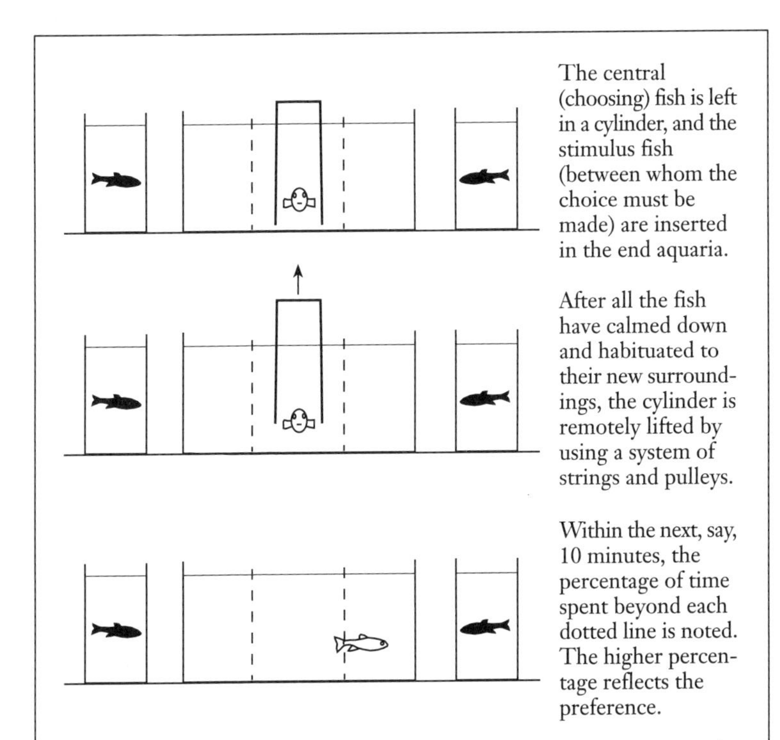

Fig. 7.3. Procedures for a visual choice test. Here the aquaria are viewed from the front. Dotted lines are drawn on the glass of the central aquarium to delineate zones of preference. The behavior of the central fish can be videotaped or viewed directly if the observer is quiet or hidden behind a tarp with slits for observation.

would be perceived as a "winner"). The third stage was a preference test similar to the one described above for the sunfish, involving the test male having to choose visually between male 1 and male 2. Out of 30 test males, 24 spent more time on the side of the loser. Females may very well assess the quality of prospective mates by comparing several of them more or less at the same time, and it seems that males may deal with that by avoiding the company of perceived Don Juans. This study speaks volumes about the subtlety of fish behavior, but in the context of this chapter we shall only note that the result is consistent with the notion that males are capable of recognizing one another.[13]

In shoaling species, individual recognition can lead to a preference for association with groups of familiar shoalmates. In visual choice tests, guppies and sticklebacks often choose to side with groups of ex-shoalmates rather than groups of strangers.[14] In olfactory choice tests, fathead minnows and blind cavefish stay closer to the water inflow that comes from a tank containing familiar conspecifics rather than unknown ones.[15] In contrast, facultative shoalers such as pumpkinseed sunfish and rock bass do not seem to exhibit strong preferences for familiar partners, possibly because they are not likely to stay together for very long in nature.[16]

What could be the advantage of keeping company with familiar shoalmates? As in Dugatkin's sunfish experiment described earlier, one possibility is that good knowledge of a shoalmate's identity and characteristics may allow a fish to stay close to those that can be bossed around. In Glasgow, Scotland, Neil Metcalfe and B. C. Thompson asked whether European minnows could recognize and prefer to associate with shoalmates that are poor competitors for food. Metcalfe and Thompson started by figuring out the pecking order within a shoal of seven marked minnows. They dispensed 100 chironomid larvae, one by one at 15-second intervals, through a pipette in the middle of the aquarium and counted the number of larvae caught by each individual. Then they captured one of the fish and divided the rest into two groups, the three most efficient food-getters on one side and the three less greedy individuals on the other. These were the subshoals offered in a visual choice test to the captured minnow (much as in Dugatkin's experiment, except that the fish had to recognize small subgroups of fish rather than single individuals). The procedure was repeated many times with other shoals. After 28 trials, the results had become obvious: the great majority of the choosing fish were spending more time near the less greedy partners.[17]

In the fathead minnow, familiarity with one's shoalmates influences antipredatory response. For example, groups of four minnows captured from the same shoal in nature show more dashing movement and aggregate more tightly in response to the appearance of a pike than do groups of four minnows that come from different shoals. Cohesive predator evasion might therefore represent another reason for wanting to hang around one's pals.[18]

Conceptually similar to familiar shoalmate recognition is kin (sibling) recognition. This is an important topic of study in animal behavior because of the related idea of kin selection, which is often used to explain the evolution of altruistic behavior (such as alarm substance emission, mentioned in chapter 1, or helping at the nest, which was presented above). Altruistic behavior presents a problem for evolutionary theorists. To take its place in the genetic baggage of a species, a trait must first spread through the population, and for this to happen, bearers of the trait have to benefit from it and produce more offspring. But the trait of altruism, by definition, handicaps its owners. If sacrificial lambs always find themselves at a disadvantage, how can their genes—including the "one" for altruism—make it to the next generation? One solution to this thorny puzzle can be found in situations in which the recipients of an altruist's largesse are its own close kin. Relatives share the altruist's genetic makeup, and so the trait can spread indirectly through them. For such situations to occur, animals must remain close to their kin, and obviously this is helped when they can recognize kin. Are fishes capable of such recognition?

Studies in the United States and in Sweden have confirmed that salmonids at least can recognize their brothers and sisters. It appears that soon after hatching, young salmon learn the collective smell of the fish surrounding them (in nature, that would be their siblings). An olfactory template takes shape within the brain of the salmon, and this template is compared with the scent of any fish encountered later in life. A match means kin; a nonmatch signifies a stranger.

The experimental evidence in support of this hypothesis is that (1) in olfactory choice tests, juvenile salmon prefer the holding water of kin over that of nonkin; (2) again in olfactory tests, juvenile salmon prefer the holding water of kin they have never seen over that of nonkin they have never seen, as long as they have had the chance to be raised with other kin that share the same family scent; (3) preference for kin disappears if the salmon are raised all alone right from birth (then they do not have a chance to form the odor template because they cannot smell any neighbors at a young age and so they cannot express a preference later in life); and (4) if the salmon are raised with nonkin from the moment they hatch, they later associate with these specific (but not any other) nonkin as much as if they were kin.[19]

Advantages to being with kin have been documented by Grant Brown and Joe Brown—no kin themselves—at Memorial University of Newfoundland. They released groups of juvenile Atlantic salmon or rainbow trout in artificial stream channels. The group members were either all kin or all nonkin. These fish established feeding territories within the channels, and the researchers watched their behavior and measured their growth. The results were that kin got along better with one another. The frequency of aggressive interactions was lower among kin than among nonkin. Kin groups also gained more weight, especially the subordinate individuals. Brown and Brown inferred other advantages such as reduced predation risk (less aggression meant fewer conspicuous movements and therefore lower detection by predators) and better overwinter survival (because of better growth and fewer injuries).[20]

Our knowledge of kin recognition is still far from complete. In all of the studies published so far, only fry or juveniles have been tested. To date, there has been no mention anywhere of kin recognition by adult fish. Perhaps adults are incapable of it, and such negative results have not been reported. Moreover, there has been no account of kin recognition based on visual cues only.

Recognition is a cognitive ability. There is a mental process by which the visual, olfactory, acoustical, or electric images of other individuals are stored within a fish's memory and associated with other characteristics such as mate status, family status, dominance status, or competitive potential. Humans can do all this among themselves without much thinking about it, but it is interesting to see that the pet fishes we keep in aquaria (and some that we don't) have similar abilities. By the way, another interesting question: Can fishes recognize individual humans? I have seen some mentions in the popular literature to the effect that they can, and some others that they cannot. I have not noticed anything in the scientific literature, although I do not pretend to have covered it all. Perhaps there is room here for aquarists to try a few of their own scientific experiments with their favorite species.

8

Gauging Predators and Adversaries

If we have a shoal of guppies, sticklebacks, or minnows in one of our fish tanks, and next to it we place a bottle containing a predatory fish (a blue acara cichlid, for example), something peculiar happens. At first the shoaling fish cower in the most distant corner from the predator, but then some of them, alone or as a small group, approach the predator in a hesitant manner, a quick lunge forward followed by a pause, then another lunge forward followed by a pause, and so on until the fish stop for the last time fairly close to the predator (less than 30 cm or so), turn away, and swim back to the rest of the shoal. What a paradoxical thing to do. Why would fish want to approach a predator?

Looking attentively at the approaching fish, we can see that they are very alert, visually fixating on the threat, beating their fins nervously, moving jerkily. For all the world it looks as if they are cautiously checking things out, assessing the danger. Well, we need look no further for an explanation; fish ethologists firmly believe that predator approach is a way for prey to take a closer look at a new fish, to determine if it is a predator, and to see if the predator looks hungry and about to attack. In fact, in many quarters the behavior is called predator inspection rather than predator approach, so certain are ethologists about the function of this action. Both freshwater and marine fishes are known to inspect predators—or any new fish for that matter—although to my knowledge only species that live in groups are prone to this behavior.

In one experiment with European minnows, shoals of 10 wild-caught fish were presented with either the realistic-looking resin model of a pike or a simple cylinder model of similar dimensions. With a system of strings and pulley, the model was made to come out of a weed clump that was located 120 cm away from a food patch in which the minnows spent their time foraging for hidden food flakes. The model approached the food patch in 10-cm steps interrupted by 30-second pauses. The minnows paid visits to both types of model; in fact they inspected the cylinder more often, an indication perhaps that they had never seen one before and were more puzzled by it. In the end, however, they must have realized that the cylinder was not a threat because they resumed foraging on the food patch even as the cylinder drew very near. In contrast, the minnows skittered more frequently after coming back from a visit to the realistic-looking pike model, and they kept their feeding levels low, which suggests that they recognized the pike as dangerous.[1]

There are many ways in which an inspecting fish could recognize the predatory nature of an intruder. First, the inspector may know what a predator looks like or smells like because it has been attacked by it in the past or because of a prior association with the presence of alarm substance or panicked shoalmates (the cases of learning we have seen in previous chapters). Second, either innately or through experience, it may recognize some of the facial features that are typical of predators, such as a large mouth and big widely separated eyes (fish often react with fear to crude dummies that display such features; see fig. 8.1).[2] And third, an inspector may treat as suspicious any action on the part of the intruder that suggests predatory intent, such as stalking or lunging. For example, if guppies are sandwiched between a tank with a satiated predator (a largemouth bass) and a tank with a bass that has been food deprived for 48 hours, they end up staying closer to the satiated bass. This is because the hungry bass at the other end is always in motion.[3] In another study, divers presented models of the predatory Atlantic trumpetfish to free-living threespot damselfish, and these prey fish reacted more strongly when the model was oriented in a strike pose.[4]

In yet one more study, researchers gave a shoal of captive European minnows a chance to inspect a northern pike isolated behind a clear partition. The minnows visited the pike at regular intervals. The pike usually kept to itself, but sometimes it could not help but try to charge at the minnows. The neat observation was that the minnows that had inspected the pike just before the final visit, that is, just before the pike broke down and attacked, seemed more alarmed upon coming back to their shoal (they fed less, shoaled more, and fin-flicked more) compared with minnows that had visited the pike a longer time prior to the attack. This suggests that minnows can read the aggressive mood of a pike. They can gauge its impending motivation to charge.[5]

Fig. 8.1. Not all fishes know instinctively what predators look like, but those that do seem to react to large mouths and big, widely separated eyes. The model faces above could be used as two-dimensional dummies presented to prey species, and, if these prey have instinctive knowledge, they should stay farther away from the model on the left than from either of the other two because it combines both scary characteristics.

Usually any rapid movement on the part of a strange species is enough to make fishes suspicious. This can turn out to be maladaptive, because fear could be elicited by any species that happen to be fairly active, even a nonpredatory one. For example, blacknose dace, which are small stream-dwelling minnows, are known to avoid areas where creek chub are present (a good idea, because chub can eat dace) and also where common suckers are present (a not-so-good idea, because suckers are totally harmless). This holds true for days when water temperature rises above 6°C, because both chub and suckers move more energetically then. At lower temperatures, chub remain active but suckers become lethargic, and the dace continue to avoid chub while now tolerating the company of suckers. It seems therefore that movement is part of the avoidance-releasing stimulus. This means that dace waste time sidestepping the innocuous suckers when these are active. But then again, mistaking the identity of a chub would be disastrous, and to safeguard against costly mistakes it may simply be safer to shun the company of any large moving fish.[6]

At first sight, inspection behavior seems to be dangerous because it brings the inspector closer to the predator. Danger may be more apparent than real, however, because some predators seem able to discern the alert disposition of inspectors and often do not bother to attack them. At Mount Allison University in Canada, Jean-Guy Godin and Scott Davis experimentally demonstrated this gauging ability of predators. They placed two aquaria side by side. One aquarium contained four guppies, and the other housed a blue acara cichlid, a natural predator of guppies in the wild. A one-way half-silvered mirror was inserted between the two tanks. Illumination could be adjusted on either side of the mirror so that in all cases the predator could see the guppies, but in only half the tests could the guppies also see the predator. When guppies could see the predator, they inspected it. In the other tests, the guppies also approached the predator but only as part of their natural wanderings because they could not see what lay behind the mirror. The behavior of the cichlid was monitored for 15 minutes under both conditions, and it turned out that although only four attacks on average were directed at the inspectors (30 different cichlids were tested), twice as many attacks were aimed at the unsuspecting non-inspectors. It seems, therefore, that the cichlids could realize the inspectors were not as vulnerable as the non-inspectors.[7] Of course a guppy still runs more risk by inspecting rather than just staying away, but this risk is less than the actual proximity between prey and predator might suggest.

In some cases, the predator may be more than just discouraged from attacking; it may be mobbed outright by the inspectors. Of course I am not referring to the actions of puny guppies in front of a mighty cichlid. Rather, what I have in mind is the reaction of aggressive butterflyfishes, damselfishes, and surgeonfishes to lone predatory moray eels or lizardfishes. After sighting one of these predators, the potential prey sometimes approaches alone or in small groups, and they lateral-display right in the face of the enemy, vigorously beating their tail toward it. Somewhat surprisingly (at least for those of us who root for predators), eels and lizardfishes often decamp after such displays without so much as a tentative lunge toward the nagging little pests. As with better-known cases of mobbing in birds, the display may encourage the predators to head for more promising hunting grounds, given that their chances of surprising a prey have suddenly taken a turn for the worse.[8]

Alert inspectors may not risk as much we might at first think, but still they do not take unnecessary chances. For example, as opposed to the pugnacious marine fishes above, small guppies do not inspect right in front of their natural predators. More often than not, they advance toward the side or the tail of the predator and wisely avoid the "attack cone" centered on the mouth area. (Interestingly, if the predator is a cichlid with a false eyespot on the tail, the guppies steer clear of the tail area as well [see fig. 8.2]; this behavior suggests that big eyes constitute danger signs and that guppies are well aware of the usual proximity of eyes to the business end of a predator—the mouth.) Guppies also give a wider berth to swimming predators than to stationary ones. Another example concerns both guppies and minnows: in tests, those individuals that originate from heavily preyed-upon

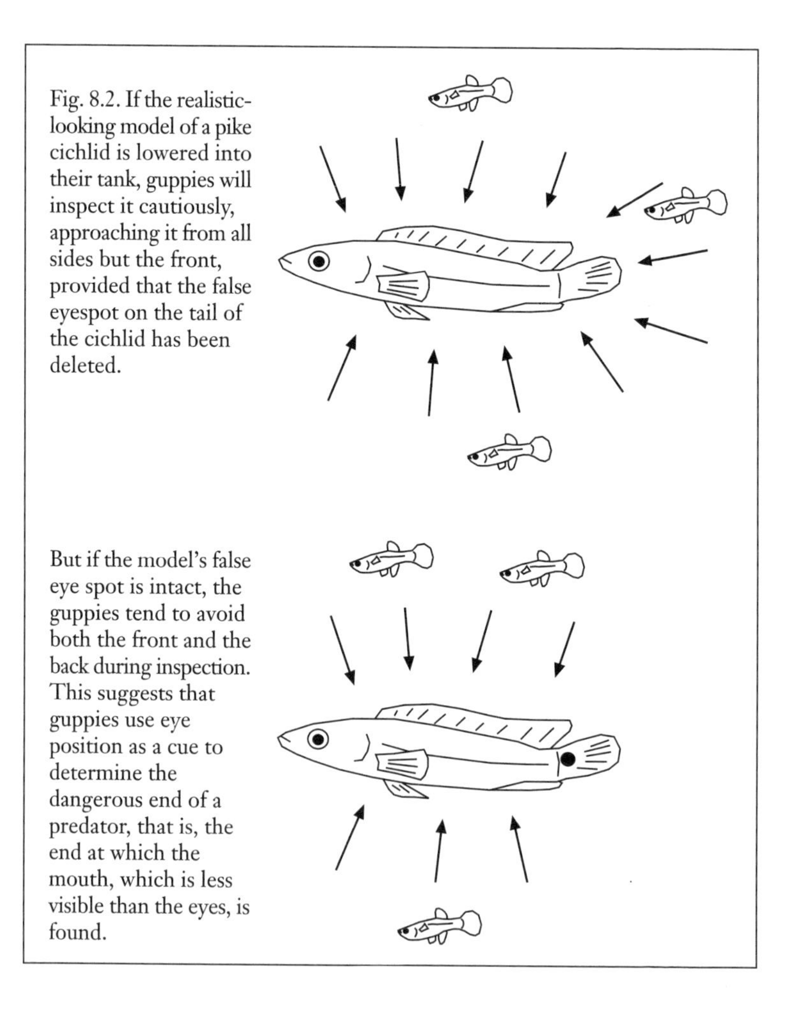

Fig. 8.2. If the realistic-looking model of a pike cichlid is lowered into their tank, guppies will inspect it cautiously, approaching it from all sides but the front, provided that the false eyespot on the tail of the cichlid has been deleted.

But if the model's false eye spot is intact, the guppies tend to avoid both the front and the back during inspection. This suggests that guppies use eye position as a cue to determine the dangerous end of a predator, that is, the end at which the mouth, which is less visible than the eyes, is found.

populations inspect predator dummies in larger groups (each individual's risk of being attacked is therefore more diluted), and they do not approach as closely as do naive fish from predator-free populations. Similarly, three-spined sticklebacks that are smaller or less armored (that is, with fewer scaly plates on their flanks) inspect predators less often and not as closely as do their better-endowed counterparts.[9]

All of the above suggests that fishes are reasonably good at gauging the risk associated with other fishes that could be predators. Now we shall see that they can also judge the strength of individuals from their own species, or more precisely, the rivals they face during fights.

Competition between fishes of the same species is a sad fact of life. Resources are seldom overabundant, and therefore animals often compete against one another for access to such vital commodities as food, shelter, and sexual partners (the latter being vital in evolutionary terms). Even when resources happen to be plentiful, quality is often uneven. Animals then fight among themselves for access to the best items. All other things being equal, animals are not willing to settle for second best; if they can, they will try every time to appropriate for themselves the biggest piece of the pie. Such competition usually takes place between individuals of the same species. After all, conspecifics are most likely to share similar tastes. For example, there is often more competition between the same-species members of a shoal grazing on algae than between two different species, even though both species may be herbivorous.

Generally speaking, such strife can take two forms. In so-called scramble (or exploitative) competition, there is no overt aggression. The most food goes to the individual that eats the fastest; the best shelter goes to the fish that reaches it first; and the most alluring female is mated by the male that finds her first or produces the most sperm. Such contests are more like races than wrestling matches. They are practiced by fish species that do not possess strong physical features for attack such as hard mouth parts, robust fins, or sharp spines. Scramble competition is also favored by gregarious species when any level of aggression would disrupt the social bonds on which the fish depend for finding food or evading predators.

The second pattern is called defense (or interference) competition, and this is where aggression comes into play. It results in fights between territory owners, chases by dominant individuals at communal food sources, or displacement in cramped shelters. Fights can be particularly interesting to watch. At first, they are often tentative: opponents may lock jaws and push or pull in a contest of strength. But if such civilized contests fail to crown a clear victor, fights get livelier and may require analysis on slow-motion videotape replays. The fish commonly circle each other quickly in an attempt to bite their opponent's tail while avoiding being bitten themselves. A few scales fly, a few fins get torn, and many muscles tire before one of the contestants finally gives up and flees from the field.

Because aggression can be so costly in terms of energy expenditure and risk of injury, it would be advantageous for fishes to be able to realize as early as possible during a fight that they are likely to get trounced and that they should quit before anything serious happens to them. This evaluation can in fact take place very early in the game, more specifically before the fight even begins. Even though they cannot see themselves in mirrors, fishes seem to be aware of their own body size and to be able to gauge the size of others relative to themselves. They can adjust their level of aggression according to the size of the opponent they are facing, or in other words, according to their probability of winning the fight. (Many studies have shown that bigger fish usually win fights.)

An experimenter can verify the size-gauging ability of fishes by presenting differently sized dummies (painted epoxy casts of dead individuals from the same species) to a territorial resident and observing its reaction. Oscars, for example, will readily charge and bite dummies that are smaller than themselves, will be more circumspect around similarly sized dummies, and will definitely hold back in front of larger ones.[10]

The golden-eyed dwarf cichlid (gold acara), however, refuses to fight larger opponents only when size difference is substantial. This aggressive fish withdraws only from opponents that are at least one and a half times heavier than itself; the rest are enthusiastically challenged. This behavior betrays either a particularly nasty disposition—not uncommon in cichlids—or poor precision in gauging relative body size.[11]

In sticklebacks, males whose throat is colored a vivid red usually win more fights. Is this because adversaries are intimidated by the red badge and back down more easily, or because they are impressed directly by the good health and fighting ability that often—but not always—go hand in hand with a red badge? At Creighton University in Nebraska, Charles Baube has used a clever trick to show that it is the badge itself that conveys information. He staged fights between males of similar size but with different amounts of red on their throat, and he did this under different types of illumination. Under normal white light (the red badge was very visible), the more colored males won more often, as expected. Under blue-purple light (the red badge appeared deep red and still contrasted with the rest of the body), the more colored males still won, showing that unusual lighting did not in itself affect their fighting behavior. However, under blue light (which rendered the badge black) or under red light (which made the whole body red and therefore hid the badge), the more colored males did not win more than about half of the fights, as we would expect from chance alone. These results can be explained by presuming that when the badge is visible, opponents are impressed by the big ones and throw in the towel more readily, but when the badge is not visible, the odds are about even because no one is unduly intimidated.[12]

Another way for fishes to size up their opponents is to observe them in action against other conspecifics. At the University of Göteberg in

Sweden, Jörgen Johnsson and Anders Akerman set up an aquarium in which two rainbow trout interacted under the watchful eye of a third individual isolated behind a partition. This observer fish was later paired either with the dominant trout it had already seen in action or with a dominant trout from another aquarium which it knew nothing about. When paired with the familiar dominant, the observer reached a decision more quickly about how to proceed with the fight. Either it gave up earlier in the fight (eventually losing) or it increased its level of aggression earlier (eventually winning). It was as if the observer had already made up its mind, from the initial viewing, whether it was stronger or weaker than the dominant. The unfamiliar contestants, for their part, were more tentative and interacted at a steadier level and for a longer period before a winner finally emerged.[13]

One of the most common ways that rivals assess each other is through ritualized displays. These are actions, sometimes exaggerated, usually performed by both contestants and specifically meant to signal fighting ability and to discourage opponents from escalating the contest (see fig. 8.3). Although aggressive displays betray a feisty mood, they are not part of a direct attack. Examples include booming sounds, water-displacing tail beats, fin spreads, gill cover spreads, head shakes, body twists, lateral displays that reveal the full size of the body, color changes, exposure of brightly colored body parts, and intricate swimming maneuvers. Fights are usually settled more quickly if the combatants have had a chance to display to each other (if only through a glass partition in the laboratory) for some time beforehand as opposed to being suddenly thrown into the ring together.[14]

The examples dealt with so far concern fish that gauge fighting ability based on body characteristics (e.g., size, color, and strength of display), but it would be a mistake to believe that physical attributes are the only determinants of the outcome of a fight. For example, if we take a fairly big stickleback that owns a territory, and we pluck it from its territory and drop it into the aquarium (territory) of another male that is smaller, we may very well see the small fish beat up the bigger one. Being in one's own territory seems to confer more confidence or perhaps a greater realization of what is at stake for the owner. Sports fans call this the home turf advantage. Ethologists prefer to speak of a prior residency effect. Within certain limits, the prior residency effect is enough to prevent large intruders from usurping the territory of smaller residents.[15]

Another factor that may influence the outcome of a fight between two closely matched fish is a prior experience of submissiveness. If a fish has just lost a contest, it is more likely to give up during the next fight as well, even though this second fight may be against a new adversary. It is as if the first setback created a general losing state of mind. In sticklebacks, this lingering effect can last for up to 6 hours. In contrast, a previous winning experience does not seem to endow the victor with greater powers.

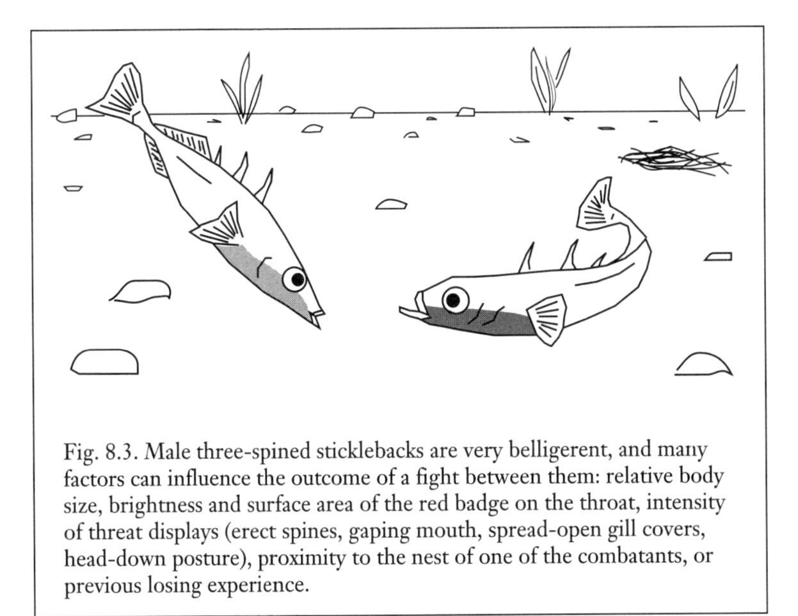

Fig. 8.3. Male three-spined sticklebacks are very belligerent, and many factors can influence the outcome of a fight between them: relative body size, brightness and surface area of the red badge on the throat, intensity of threat displays (erect spines, gaping mouth, spread-open gill covers, head-down posture), proximity to the nest of one of the combatants, or previous losing experience.

As an example, we can look at the work of Theo Bakker and his colleagues at the University of Leiden in the Netherlands. These researchers kept a great number of territorial male sticklebacks in individual tanks. They chose a few individuals at random and subjected them to a losing experience by dropping them into the tank of another male; this other male quickly established dominance, taking advantage of the prior residency effect because he was in his home territory. Other males were also chosen at random, and they experienced a win by having another male dropped into their own tank (for them, the prior residency effect worked in their favor). Three or six hours later, these respective losers and winners each met an inexperienced male within the confines of a neutral arena unfamiliar to all of them. If it were not for their previous experience, we would expect the previous losers as well as the previous winners to dominate this new encounter on only half of the tests because all of these fish were chosen at random. Previous winners indeed won only half of the time, indicating that their previous winning experience did not make them stronger. But a different picture emerged in the case of the previous losers: none of them won a single fight when the encounter was held 3 hours after their first debacle. Even after 6 hours, the previous losers won only 20% of all tests.[16]

Similar results have been obtained with blue gourami, paradise fish, green swordtail, and pumpkinseed sunfish.[17] Perhaps social experience can change the endocrine baggage of an animal, and although we still do not know exactly what those changes are in feuding fishes, they seem to make a difference only after experiencing a defeat.

Recently, researchers have started looking at other determinants of dominance during fights, determinants that are more difficult to measure than body size, prior residency, or prior experience. Two such factors are stamina and motivation. Escalated fights may last for a long time (a half hour is not uncommon in some species), and stamina would prove an asset in such a situation. Let's consider the work of

Francis Neat and co-workers on the redbelly tilapia. These researchers found that losers at the end of territorial fights harbored more lactate within their muscle than winners did. Lactate is a metabolic by-product that can cause fatigue, so the vanquished fish may have lost because they were the first ones to tire.[18]

Motivation, or fighting spirit, could also characterize good combatants. In another study on redbelly tilapia by Neat, small males sometimes won in fights with larger ones. Careful analysis of video-recorded bouts showed that these smaller winners were more aggressive during the fight and inflicted more bites. They also had larger gonads, which may indicate that they were closer to spawning time and that the territory they defended was therefore perceived as more valuable.[19]

Remembering Who Is the Boss

Once a fish has fought and lost against a superior opponent, does it simply learn once and for all who is the boss, or does it keep on reassessing the fighting ability of that winner as time goes by? Researchers have found examples of both behaviors. At Simon Fraser University in Vancouver, Jeremy Abbott and his co-workers illustrated the permanency of dominance status through the following experiment. They let two rainbow trout establish a dominance relationship between themselves. If one fish was 5% larger than the other, it always won. Then the researchers separated the two fish and fed the subordinate in excess, so much so that eventually it became at least 15% bigger than the dominant. The two fish were reunited, and surprisingly the subordinate still cowered in front of the dominant despite its newly acquired size advantage. Abbott and his colleagues concluded that risk of injury during fighting is so pronounced that trout prefer to use memory rather than renewed combat to settle contests between themselves.[a]

On the other hand, John Todd has reported that catfish may continuously monitor dominance status by olfactory means. Todd forced a pair of yellow bullheads to share a 190-liter aquarium. One member of the pair was clearly dominant over the other, forcing it to flee at every encounter. When this dominant was removed, kept in a separate aquarium overnight, and then returned, the submissive apparently recognized it right away and fled from it again without further ceremony. However, if during its overnight leave of absence the dominant bullhead was exposed to, and beaten by, an even more dominant catfish, then upon the return of that former bully to the home tank, the submissive individual attacked it. Todd and his co-workers interpreted this incident as a change in the chemical label of the removed fish, which was brought about by its defeat during the overnight contest and was recognized by the submissive as a change in status. Maybe catfish can smell fear.[b]

People are familiar with the concept of body language. We apply it in the context of our interpersonal relationships, and we even fancy that we can read it in many of our mammalian pet companions and livestock. But who would seriously talk of a body language in fishes? To the casual observer, fishes seem too placid and expressionless. Yet for the fishes themselves, there must be such a thing as body language, and they must be well versed in it because they seem to be able to read the aggressive mood of predators and rivals and to translate this knowledge into an evaluation of their risk of being eaten or beaten. This body language is based on a syntax of size, color, odor, and movements.

Part 3

Choosing Abilities

9

Mate Choice

Almost all fishes reproduce sexually. During the breeding season, most mature individuals seek out reproductive partners. In many species, the pairing up of males and females does not occur at random. Some individuals are accepted as sexual partners whereas others are rejected. In polygynous species, some males sire broods from several females while the nests of other males remain empty. This is mate choice. Many fishes have the ability to distinguish between potential partners of different quality and to choose in a way that makes adaptive sense. In this chapter, we explore the factors that make a fish sexy in the eyes of a prospective partner of the other sex.

But first, two digressions. The first one is that not all fishes exhibit mate choice. For example, in species characterized by a solitary lifestyle, encounters between males and females may represent a rare event, perhaps the only chance either one of them will ever have to mate. This is a chance not to be squandered through finicky tastes. So mate choice is absent in those species. Deep-sea anglerfishes (the ceratioid family) represent an extreme example of this. Females lead a sedentary life at great depths. Males are much smaller than females. Such dwarf males are well known for their habit of attaching themselves to a female after the two happen to meet. Dwarf males remain fixed to their mate in this way for the rest of their lives (in some species), sometimes even developing a parasitic connection between their blood vessels and those of the female. In this unorthodox way they will survive, nursed by the female and the nutrients her blood carries, and remain on hand to pass their sperm to the female when the time comes. Very nice, but there seems to be little room here for mate choice.

At the other extreme, in species that gather in huge shoals to breed (herring and cod, for example), mate choice may also be lacking. These species indulge in group spawning, a synchronized event during which all males and females shed sperm and eggs simultaneously in the same locale. Because sperm and eggs get mixed up at the whim of local water currents, there is theoretically no mate choice. I say "theoretically" because we must concede the possibility that a given male–female pair may stay close to each other during a group spawning, thereby increasing the chances that their gametes will meet and fertilize one another. Unfortunately, this is hard to prove: imagine following a particular pair of herring in a throng of thousands, and you will quickly see the difficulty.

In species in which group spawning is the norm, males are often endowed with testes many times bigger than those of males in species that enjoy more private matings. This makes adaptive sense: in group spawning, the sperm from each male are engaged in a race with those of all the other donors present. So the more sperm a male releases, the better his chances of fertilizing many eggs and leaving a disproportionately high number of descendants in the next generation. This is called sperm competition. Sperm competition drives males to invest a lot of energy in producing sperm, and this is reflected in the anatomy of their reproductive system, namely, the impressive size of their gonads.[1]

My second digression is about who chooses whom. In general, females are choosier than males. This is because the production of an egg, with its self-contained food store to provide for the development of the embryo, is a more costly investment for the female than production of tiny spermatozoids—however numerous—is for the male. For a male, sperm are relatively cheap—relative, that is, to eggs. (Sperm production is not entirely without costs, however, otherwise all males, not just group spawners, would have big gonads.) Because their sperm are relatively cheap, males often do not mind dispensing sperm onto almost any receptive female, sexy or not, that comes their way. A female, on the other hand, does not necessarily benefit if her precious eggs are fertilized by any nondescript sperm donor. She is keen to choose the very best father for the offspring into which she has already invested so much. So, the study of mate choice is often the study of those characteristics that make males sexy in the eyes of females.

There are exceptions, however. Sometimes males are choosy too. This happens when males invest more than just sperm into reproduction. In the majority of fish species in which parental care is present, it is the male alone that tends to the eggs. His investment is behavioral: he is the one that builds or cleans the nest, defends the territory around it, and attacks potential egg and fry predators. His investment is also substantial, both in terms of risk to life and limb and in terms of energy expenditure. The male would like all of this effort to be devoted to a batch of eggs that is large (giving him many descendants) and of good quality (increasing the chances of fry survivorship). Accordingly, he preferentially mates with fit females that can deliver large batches of big eggs to his care.

Are Sperm That Cheap?

Sperm may be cheaper to produce than eggs, but this does not mean the cost of their manufacture is negligible. After all, males do not broadcast sperm indiscriminately in the environment. In fact, some males can be downright stingy with their sperm. The first thing to know here is that male fishes seem to be able to control the number of sperm (the size of the ejaculate) they release during a spawning event. The second thing is that they appear to exert this control in a way that makes sense depending on the circumstances. In some coral reef species, for example, a male may need to spawn with many females, sometimes in the same day, and it has been found that males do not apportion sperm equally between all females. With larger females, which usually produce more eggs, males release more sperm. With smaller females, males are less generous. The males do not deign to release more sperm than are necessary for a reasonable fertilization rate, keeping the rest in reserve for future spawnings. Biologists call this behavior sperm economy.[a]

Another example of sperm economy can be found in species in which a male defends a territory that encompasses the domain of many females. These females become the male's harem. It has been found that the larger a male's harem, the fewer sperm he releases when spawning with any single female. He wants to make sure that he has enough sperm to go around the full harem, and this means smaller individual contributions when the females are more numerous.[b]

Anyone wondering how researchers manage to obtain information about sperm allocation should prepare to be surprised at how simple it is. Snorkelers hover above likely spawning sites, making sure that they do not disturb spawning pairs. But as soon as the deed is done, the snorkelers descend, scatter the fish, and trap all of the milt and eggs—still visible as a white plume in the water and still fairly localized—within a large plastic bag. This water is brought back to a boat, mixed, and sampled. The eggs and sperm are stained and counted under a microscope. The tally is multiplied by the dilution factor of the sampling to obtain a final count.

So what are the criteria of mate choice? What defines a sexy partner? The methodology used to answer these questions should by now be familiar to the reader: it is the choice test. A fish—say, a female in reproductive condition—is placed in a central aquarium sandwiched between two other tanks that each contains a male. A note is made of how much time the female spends next to each tank. It is assumed that she spends more time next to the male with which she would prefer to mate. This assumption can be verified by placing a spawning substrate on each side of the central aquarium and observing that the female lays eggs on the side at which she had spent more time during the choice test.

It is also possible to allow the female to interact directly with both males and spawn with one of them, but this usually means that the two males can also interact with each other, and the dominant male may actively preclude the other one from approaching the female and presenting his case. This male–male interaction is an example of *intra*-sexual selection, and although this behavior has a bearing on mating in nature, it messes up the picture we would like to draw about mate choice—a behavior that corresponds to *inter*sexual selection. So we usually try to eliminate the influence of male–male interactions during experimental protocols.

And now, onward with the list of sexy characters in fishes.

In amorous fishes, size does matter. Body size, that is. Males prefer larger females, probably because big females can produce more eggs, or at least bigger eggs that hatch into bigger offspring, which may have a better chance of survival. Females also fancy larger partners, probably because big males can better defend nest and progeny if such is the pattern of the species. Moreover, for both sexes, large size can only be attained if the fish feed well; therefore, being big is a sign of good foraging ability, a good trait to pass on to one's progeny. Large size is also correlated with greater age (fishes are indeterminate growers, meaning that they keep growing throughout their life), and so large size is an indicator of good survival ability, another good trait to pass on to one's offspring.[2]

At Indiana University, Bill Rowland often teased his sticklebacks by presenting them with dummies during the reproductive season. In one experiment, he dangled fake males in front of a female and moved those dummies in ways that simulated a courtship dance. Rowland observed that the female preferred to follow the largest male. Rowland also presented two female dummies with extended bellies to a male and saw that the male courted the largest female more assiduously. What was even more astounding is that when one of the dummies was grossly oversized (e.g., a male 25% bigger than the largest stud in the original population or a female with a belly distended well beyond normal limits), this exceptional dummy was still preferred. This is an example of what ethologists call a supernormal stimulus.[3]

Preference for larger partners does not mean that small fish do not get to breed. They simply pair up with other small partners. In their laboratory at the University of Western Ontario, Miles Keenleyside and his students have used standard choice tests to show that in convict cichlids, both males and females prefer larger mates. Then Keenleyside placed convicts of both sexes into a large outdoor pond in summer. He observed that large males and females paired up together, as did small males and females. This pattern is called assortative mating, the seeking of a partner that is similar to oneself (although assortative mating is not necessarily by choice in the case of the smaller individuals here).[4] Keenleyside noted that pairs of big as well as pairs of small fish reared their young successfully. Offspring survival was not measured, however, and we may surmise that offspring from large parents were more likely to survive longer. Another study from a different laboratory and with another cichlid, the black-belt, found that the young from large parents grew faster and survived longer.[5]

During his Ph.D. studies undertaken in Mart Gross's laboratory at the University of Toronto, John Reynolds carefully monitored the individual mating behavior and reproductive success of a great number of guppies he kept in a battery of aquaria. He reported that females were attracted to large males, that larger males produced larger sons and faster-growing daughters, and that these daughters in turn produced more eggs when they reached adult age. Thus Reynolds demonstrated that female preference for larger males is adaptive because it leads to better and sexier offspring in the next generation and to more descendants in the generation after that.[6]

Most choice tests involve the simultaneous assessment of mate size, or more plainly, the fact that a fish must choose between two or more partners that can be seen, and therefore directly compared, more or less at the same time. What would happen if the assessment had to be sequential, if potential mates had to be compared one after the other? When a female views a lone male, she must bear in mind that the previous candidate has been left behind and that the next candidate is of unknown quality. The best moment to spawn may be here and now, while interacting with the potential mate that happens to be present. Yet making a choice on the spot is a gamble because the female does not know if the next candidate will turn out to be better, nor does she know whether she will be able to find the present candidate again.

Two rules of thumb could conceivably be followed by the choosing fish. One is "I'm going to mate with the first individual I meet that satisfies a minimum threshold criterion for size, such as anyone bigger than my own body size, for example." The other rule of thumb could be "I'm going to evaluate the individual's size based mostly on what I have seen before and accept to spawn if this present individual represents an improvement." An experiment on mottled sculpins has provided evidence for this latter tactic. Small, medium, and large males were allowed to establish territories in individual aquaria. A gravid female was slipped into each tank and left there for 24 hours. If the female did not spawn with the male, she was removed and inserted into the tank of a differently sized male. On this second attempt, the chances that spawning would occur were twice as high when the transfer had been from small to medium male rather than from large to medium male, lending credence to the notion that female sculpins use a relative criterion to assess male size. Their preference was influenced by what they had seen before.[7]

Other criteria for mate choice are the presence and quality of nuptial ornaments on the body of the chosen fish. Although there is at least one interesting exception (we return to it shortly), nuptial ornaments seem to develop mostly in males for the specific purpose of wooing females. Three species in particular have drawn the attention of ethologists: three-spined sticklebacks with their red throat, guppies with their orange spots, and swordtails with their lower tail extension.

A great number of researchers have established that female three-spined sticklebacks prefer males with throats that are more intensely red, or at least any color that contrasts well with the rest of the body. A bright red throat indicates a healthy parasite-free condition as well as high success in territorial fights and parental duties, and this could justify the females' choice.[8]

In guppies, females generally prefer males that have more and brighter orange spots on their flanks. This is a true preference for the spots themselves and not for some correlated physical attribute. We know this because preference for males with bright spots disappears when choice tests are conducted under orange light, which hides the spots by making the whole body appear uniformly orange without affecting any other physical attributes.[9] Moreover, males that are fed carotenoid supplements—carotenoids are the pigments that provide the orange color in spots—are preferred over normally fed males, even though the carotenoid-fed males' body size and courtship behavior are not intensified, only the brightness of their spots.[10]

There is evidence that males that are naturally more orange are also in better physical shape. When Paul Nicoletto, then a graduate student at the University of New Mexico in Albuquerque, placed brightly colored males in a flow chamber, he found that they were able to swim in a strong current for longer periods of time than were their drab brethren. By choosing brightly colored males, female guppies may be favoring physically fit partners.[11] Moreover, as in sticklebacks, brightly colored males tend to have fewer parasites.[12]

An exception to this preference for brightly colored male guppies comes from populations in Trinidad, where predators are abundant. There, colored males may be fit, but they are also more visible to predators. Therefore, orange spots on a prospective mate transmit an ambivalent message to the females: they announce a physically fit male but also one whose sons will inherit bright colors and therefore run a greater risk of being picked up by predators. Not surprisingly, females in these populations express only a lukewarm preference (and sometimes outright rejection) for brightly colored males. Moreover, when many populations are compared, there is a good correlation between the degree of female preference for bright males and the average brightness of all males in the population. In populations in which female preference is strong, males end up being more colorful. So it seems that male brightness is a trait whose evolution is driven by female choice in guppies, and female choice in turn may depend on predation pressure.[13]

Swordtails are aptly named fishes. Males in these species possess an elongated lower tail section that resembles a sword. Just like brightly colored spots, such a bizarre and extravagant body part seems to fulfill no function other than to impress females. Charles Darwin himself had suggested that the sword of swordtails enhanced the attractiveness of males to females. Do female swordtails choose males with longer tails in the same way that female guppies choose males with bigger and brighter orange spots? The answer is yes. Working at the University of Texas at Austin, Alexandra Basolo offered a choice of two males with different tail lengths to female green swordtails. When presented with males whose tail length was 69 versus 45 mm, females spent twice as much time in front of the longer-tailed male. Was it because that male had other characteristics aside from his sword that made him attractive? The answer here is no. If the 69-mm tail was shortened with a scalpel down to 24 mm and the 45-mm tail on the other male was cut down to 36 mm so that the previously shorter-tailed male was now the longer-tailed one, the females reversed their preference and spent nearly twice as much time in front of the male that was now blessed with the longest ornament (see fig. 9.1). So tail length is indeed an important criterion for mate choice in female swordtails.[14]

Earlier I alluded to an exception to the preponderance of male ornaments. This exception can be found in pipefishes. In these fishes (as in the closely related seahorses), males provide extensive parental care by incubating eggs within a special brood pouch on their belly. Room within the brood pouch of the best (largest) males is limited, and females must therefore compete for such males. Accordingly, in many species, females have developed nuptial ornaments to embellish themselves and improve their chances of being accepted by a desirable male. They exhibit special colors, bar patterns, or skin folds on the body. Several experiments have indicated that males mate preferentially with females that better display such features.[15]

A new idea making headlines in the field of mate choice is the notion that animals with symmetrical body features should be more attractive to prospective partners. By symmetry I mean that the right side of the body mirrors the left side as closely as possible. The rationale behind this concept is that symmetrical features can only be the outcome of a smooth developmental process, unperturbed by misfortunes such as a temporary lack of food, disease, or exposure to pollution. The capacity to develop evenly could also be inherited, the result of having "good genes." Symmetry might therefore indicate a capacity to avoid detrimental conditions and to perform well physically. All of this leads us to believe that symmetrical partners could be better parents, better defenders of the progeny, and in the case of females, better egg producers (that is, they could produce more eggs). Symmetry as a basis for mate choice was first tested and confirmed in insects and birds. Recently some studies have also involved fishes.

Males of the swordtail *Xiphophorus cortezi* bear a variable number of vertical dark bars on their flanks. Using a freeze-branding technique

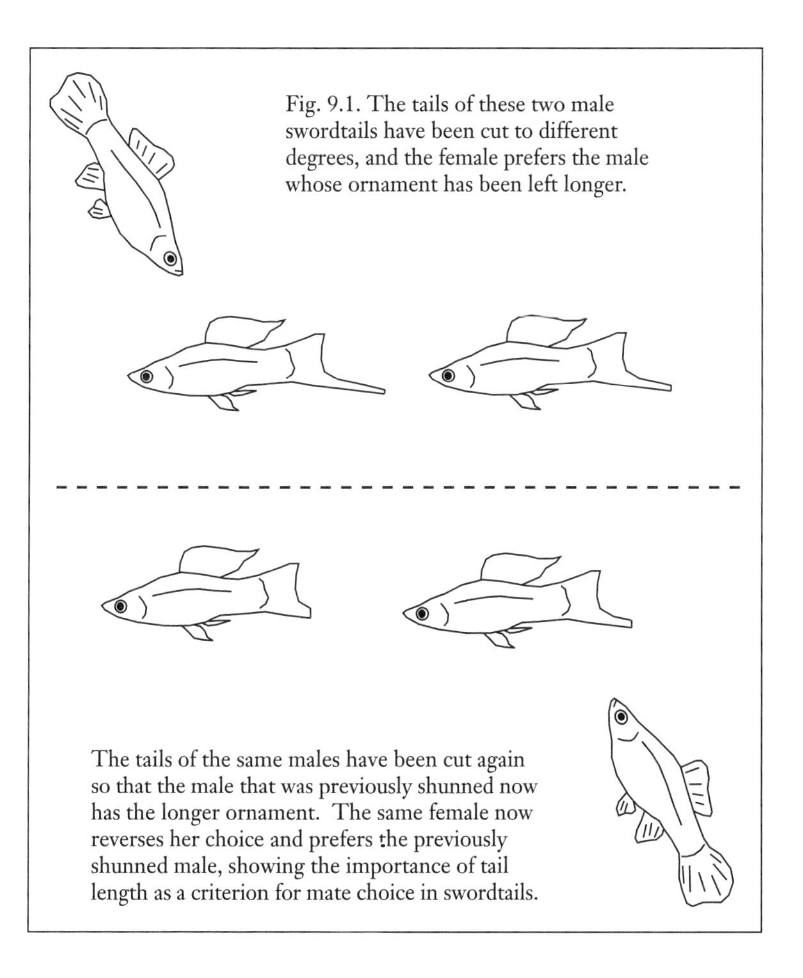

Fig. 9.1. The tails of these two male swordtails have been cut to different degrees, and the female prefers the male whose ornament has been left longer.

The tails of the same males have been cut again so that the male that was previously shunned now has the longer ornament. The same female now reverses her choice and prefers the previously shunned male, showing the importance of tail length as a criterion for mate choice in swordtails.

that erases some of these bars, Molly Morris and Kenneth Casey from Montgomery College in Maryland and from the University of Texas at Austin created males that were either symmetrical or asymmetrical in the number of bars they sported on each side. When given a choice between a male that had 6 and 6 bars versus one that had 5 and 7 (note that in both cases the total number of bars per fish was the same, 12), females always spent at least twice as much time near the symmetrical male. Even more convincingly, if the condition of these males was switched around (for example, the 6 / 6 male was made to be 4 / 6 by erasing 2 bars on one side, whereas the 5 / 7 was made to be 5 / 5 by erasing 2 of his bars as well; note that the resulting total number of bars is still the same for both fish), the females reversed their choice and then spent more time near the newly symmetrical male. This is a convincing demonstration that female *X. cortezi* prefer symmetrical males. Similar conclusions have been reached in experiments with other flank-marked species such as the sailfin molly and the guppy.[16]

All mate-choice criteria mentioned so far have had to do with physical attributes, but behavior can also be involved in mate choice. When in breeding season, fishes are keen to show that they are ready to mate. To this end, they often send multiple signals. Not only do they develop nuptial ornaments on their body, but many of them also display courtship behavior. They can adopt specific postures or execute elaborate maneuvers. Not surprisingly, fishes that seek sexual partners tend to prefer those that display over those that do not. In sticklebacks, for example, males will spend more time with the dummy of a ripe female that is presented at an angle (head up, the typical courtship posture of females) over one that is presented horizontally.[17] If the male is to hold any hope of wooing a female back to his nest, his first task is to ascertain that the female is amenable to being wooed, hence his preference for females that send appropriate signals.

The same argument holds for movement instead of posture. A fish that executes courtship moves is more attractive to a prospective mate than one that shows, say, feeding behavior. While working at the University of California at Davis, Gil Rosenthal and his colleagues gave female green swordtails the opportunity to watch television. For the viewing pleasure of this piscine audience, the television offered a video-taped sequence of a single male green swordtail that was either actively feeding or amorously courting; the object of his affection was another female that, on the video, had been digitally erased. Although the male was the same in both sequences, and although the intensity of his movements was comparable in both courting and feeding, the females that watched these two shows did not respond equally to both. They faced the television, motionless (glued to the screen), more often when the courting male was on.[18]

We can ask whether the rate of courtship display (not simply its presence) can influence mate choice. Would a female, for example, prefer a male that displayed a courtship maneuver 10 times a minute as opposed to a more half-hearted 5 times a minute? In guppies at least, display rate seems to be very important indeed. It may even mitigate the effect of body size. Bigger males are normally preferred by females, but if the display rate of big males falls below that of the smaller guys, females then abate their normal preference and dispense their favor on both types of male equally.[19]

It is noteworthy that female guppies that come from predator-rich environments do not exhibit this normal preference for actively courting males, in the same way and for the same reason that they shun brightly colored ones: the sons of such males would stand a greater chance of falling victim to predation because their excited demeanor renders them more visible.[20]

In guppies, a typical courtship maneuver is the so-called sigmoid display, in which males flex their body into an S shape and vibrate rapidly. Paul Nicoletto showed that the duration of this display is important in the eyes of females and that long displays can only be sustained by males that are literally better built. Nicoletto raised male guppies in two kinds of aquaria, one with high water velocity and one with little current. Food was plentiful in both types of environments, and so the fish grew well. However, the fish from the high-flow environment also developed bigger muscles from all the extra swimming they had to do (this is the only experiment that I am aware of in which fish were sent to a gym for body building). These hunks turned out to be more adept at displaying to females for a long time, and in choice tests they were indeed preferred by females. Because these fish could also swim faster, as measured by Nicoletto in experimental flow chambers, we can conclude that through their preference for long displays, females were selecting for males that were more fit physically.[21]

As with nuptial colors, display rate could also be viewed by females as an indicator of parasitism in males. Parasitized male guppies perform the sigmoid display less often than do healthy ones, and they are disregarded by females in choice tests.[22]

Field observations of bicolor damselfish on a reef off St. Croix, U.S. Virgin Islands, revealed that females did not mate preferentially with larger males. Instead, their choice was based solely on male courtship rate. Those males that performed more dips (a courtship behavior) per unit time ended up with more eggs inside their nest. Males guard eggs for approximately 2 weeks in this species, and egg survival turned out to be more likely in the nest of those males that had "dipped" more. This finding strongly suggested to the biologists who had conducted this study that courtship rate was an honest indicator of parental quality and dedication. In bicolor damselfish, it seems that Don Juans are also good fathers.[23]

In species whose eggs develop under the guardianship of an attentive father, the quality of the territory held by the male or the nest site he has cleaned or built may become an important factor for mate choice. Females may base their preference on territory characteristics rather than, or in addition to, male attributes.

In one experiment, Craig Sargent kept four male three-spined sticklebacks in a circular wading pool. He placed four patches of plastic seaweed 90° apart along the periphery, and each male ended up building a nest next to a seaweed patch. Then Sargent removed the seaweed and covered two of the nests with a half flowerpot so that these nests remained accessible but were concealed from the prying eyes of passersby. Then one by one he slipped 13 females into the pool and noticed which nest each female chose. When she began to enter a nest, Sargent rudely interrupted and removed her so that she did not lay eggs in the nest, something that might have influenced the choice of the female coming after her. After all females had made their choices, Sargent uncovered the two concealed nests and placed the flowerpots over the other two nests. Thirteen more females were asked to choose between the various nests. This experiment was repeated with different fish in two other wading pools, for an overall total of 12 males tested both with and without a flowerpot over their nest. The results were that for 11 of those 12 males, the nest was chosen more often when it was covered than when it was exposed. It seems therefore that female sticklebacks prefer to spawn in nests that are better concealed, probably because such nests are less likely to be raided by predators.[24]

In nature, males seem to be very much aware of the kind of spawning sites that females like, and so there tends to be a correlation between male quality and territory quality. Big males, which are more competitive and are already preferred by females, also tend to hold the best territories. Examples include big mottled sculpins defending the largest tiles under which females like to spawn, the most colorful pupfish holding territories with more nooks and crannies into which eggs can be deposited, and big sand gobies gaining access to a limited number of nests and getting all of the females' attention.[25] This makes it difficult for experimenters to tease apart the respective effects of physical attributes and territory quality on female choice.

In Craig Sargent's work with the stickleback, as described above, this difficulty was circumvented by manipulating territory quality independent of male quality, something that is easier to do in the laboratory. In another example, this time from the field, Sue Thompson studied the reproductive behavior of mottled triplefins in New Zealand. On shallow rocky reefs, males defend territories that are approximately 2 meters square and that usually encompass small rocks or large boulders. Females prefer to mate within territories that contain more rocks (more potential for hiding the eggs) and that are closer to large boulders (more protection during storms). Big competitive males secure such territories for themselves and enjoy a greater reproductive success, as measured by the number of eggs that can be found within their domain. For their part, small males on relatively barren grounds are completely overlooked by females.

In her experiment, Thompson removed rocks from the territory of the top male (which had the most eggs), added rocks to the home of the most subordinate male, and left untouched the territories of the 10 males in between. Because theirs is a long reproductive season, the males kept courting females during the following month, and Thompson was able to monitor their success (the number of eggs within each territory). The untouched territories did not show a great change in egg number, but the small male, with his enriched territory, now became one of the most successful mates. He greatly outdid the success of the big male, whose mating success, in his now-impoverished territory, took a nosedive.

This result indicates that territory quality is important, but it does not mean that male size holds absolutely no value in the eyes of females. Thompson performed another experiment in which she removed the top male from his territory. A small male from a barren territory moved in to replace him (the other males were content with their own real estate). Thompson observed that although this small male did better in his new home than he had before, he still did not gather as many eggs as the big male had in the same spot. From this and the previous experiment, Thompson concluded that both territory quality and male size influence the mating decision of female triplefins.[26]

In contrast, for the bluehead wrasse, territory quality is paramount in the eyes of females. There is no parental care in this reef species. The eggs are shed, fertilized, and left to the mercy of the currents. However, females prefer to spawn over downcurrent segments of the reef, and big males establish spawning territories there. If a particularly successful male is experimentally removed from his territory, females that used to frequent his site still go there and mate with whichever male happens to take over. If this new male is a neighbor that has decided to move upscale and trade up his old site for a new one, the females that used to visit his old site do not follow him. They remain faithful to the old site. This suggests that male quality is not important; only preference for a given territory counts.

So, in bluehead wrasse, the females are the ones that decide what constitutes a good spawning site, and once they make up their mind, they stick to their decision. If those females are experimentally removed and replaced by other females from a different reef, the new arrivals may disregard the traditional spawning sites—over which the resident big males still hover—and hang around a new and empty spot. It takes a few days for an astute big male to give up on his territory and claim ownership of this new hot spot, where he can resume mating. Such field experimentation and observations were conducted by Robert Warner, working on a reef off the Caribbean coast of Panama, near the field station of the Smithsonian Tropical Research Institute.[27]

One last interesting example on the role of territory quality: in the bicolor damselfish, females may evaluate the quality of a territory by monitoring the success of previous broods. Females of this species mate repeatedly throughout the reproductive season. At a field site in the U.S. Virgin Islands, Roland Knapp observed that although females normally show good nest fidelity from one spawning event to the next, they switch to other nests—and therefore to other males—if one of their previous broods happens to fall prey to nocturnally feeding brittlestars. Females are able to detect the lingering smell of brittlestars and to see that their eggs have been destroyed, and during subsequent spawnings they avoid the nest in which the tragedy took place. This is a good adjustment because brittlestars, like lightning, seem to strike more than once in the same place, as evidenced by the fact that new broods usually do not survive well in those nests that have been depredated in the past.[28]

Territory quality is not the only extrinsic parameter that may influence the sexiness of a male. In the early 1990s, Lee Dugatkin provided the first evidence that mate choice by females may be influenced by the sight of another female making her own choice. Dugatkin allowed a female guppy to watch two males, one each in adjacent aquaria on either side of her. One of these males was alone while the other one was busy courting a female. Later, the central female was given a choice between these two males, both of which were now alone. A total of 20 females were tested in this way. Seventeen of the 20 females eventually spent more time next to the male that they had seen with the other female. Dugatkin concluded that under certain conditions, females tend to copy the choice of other females in the population.[29]

The qualifier "under certain conditions" is important. Subsequent research showed that mate-choice copying is not practiced by pet shop guppies; only fish captured in the wild exhibit the phenomenon, and even then not all populations do. Neither is it expressed when a predator is shown to the choosing females or when the females are hungry: they choose randomly then, an indication perhaps that they are not concentrating really well on the task at hand. Moreover, whereas young female guppies tend to copy the choice of older fish, the reverse does not hold true. Finally, other species such as sticklebacks have not yielded any evidence of mate-choice copying when tested.[30]

In studies of mate-choice copying, it is important to disprove alternate explanations for the observed female behavior. For example, we could conceive that an encounter with a female could "prime" the lucky male and cause him to become more ardent in his subsequent courting of the central female. The preference of the central female for this male could therefore be a function of his courtship intensity rather than an attempt at copying. But if that were the case, we could swap the central female for a naive female from another aquarium and predict that this new female would also choose the presumably primed male. Dugatkin tried this in his original study. Of the 20 naive females he tested, only 11 sidled up to the primed male, whereas 9 chose the male that had been alone, a result that did not differ from random choice. Dugatkin concluded that priming of males does not take place or that it cannot be perceived by females. That left mate-choice copying as the best hypothesis to explain the behavior of the female guppies.

A variant test of the priming hypothesis was conducted by Ingo Schlupp while he worked on sailfin mollies in Michael Ryan's lab at the University of Texas in Austin. In those experiments, a central female could see a male on either side of her. Both of these males were interacting with a female through a transparent glass, but only one of those side females could be seen by the central one thanks to the judicious placement of an opaque partition (see fig. 9.2). The central female subsequently spent more time close to the male that had been witnessed in the presence of another female. There is no reason to believe that the behavior of the two males was different (their previous experience had been exactly the same), and therefore choice must have been based solely on the view of the other female.[31]

Another form of mate-choice copying is the tendency by some females to spawn in nests that already contain eggs. So far, this preference has been reported for at least 13 species of freshwater and marine fishes.[32] The difficulty for the experimenter here is to show that egg presence exerts a direct influence and is not simply a correlate of some other, more important factor. If, for example, all females in a population prefer to spawn with the male that courts the most, then obviously many females in a row will spawn with him, which means that all females but the first one will spawn in a nest that already contains eggs. This does not prove a preference for nests with eggs but simply a choice for a given male that courts a lot.

This problem can be tackled in various ways. One is to document the characteristics of males with and without eggs in their nest and to show that they are not different. Another is to show that females do not seem to care which male they choose unless they have a chance to visit his nest. Both methods have been applied to a population of sticklebacks, and the concept of preference for nests with eggs was sup-

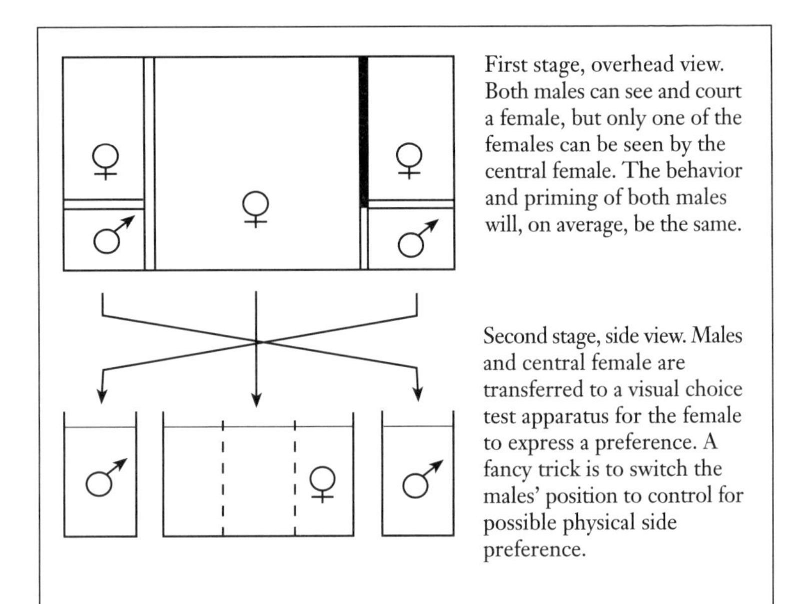

Fig. 9.2. Experimental setup for testing the idea of mate-choice copying while controlling for a possible priming effect on the males.

ported.[33] Another way is to exchange nests with and without broods between males and to observe a consequent switch in female preference, a result that excludes the possible influence of body characteristics (however, the behavior of the males may still change, depending on their ownership of eggs, making it necessary to measure whether such a change occurs).[34] Yet another method is to place a male in a bottle midway between two nests, one with and one without eggs, and to show that females prefer to lay their eggs in the already-occupied nest. Such an elegant experiment was part of Sarah Kraak's Ph.D. thesis at the University of Groningen; it demonstrated that females of the Mediterranean blenny *Aidablennius sphynx* do indeed base their choice on the presence of eggs in the nest.[35]

There are subtleties in this preference for nests with eggs. In sticklebacks, for example, females prefer to spawn in nests that already contain two or three clutches, but they disregard nests that are packed with four or five.[36] This may be related to the fact that eggs in an overcrowded nest have trouble getting enough oxygen for optimal development. In the same vein, female common gobies normally prefer nests that already have eggs in residence, but if the fish are kept in oxygen-poor water, the females reverse their preference and favor nests that are empty. This new choice is probably adaptive because competition for oxygen is fiercer in a nest with many eggs, and such a nest should therefore be avoided when oxygen is already in short supply.[37]

Female preference for nests with eggs does not leave males indifferent. In fathead minnows, for example, big newly reproductive males sometimes evict the owner of an already established nest rather than occupy a similar but empty site. Such usurpers do not destroy the previous owner's eggs but instead care for them. Why do these males care for eggs that are not their own? The behavior in fact makes sense when we learn that female fatheads, like other species, prefer to mate with males that are already caring for eggs. This preference may very well have led to the evolution of nest takeovers and adoption of eggs.[38]

In three-spined sticklebacks, territorial males have sometimes been observed dashing over to the nest of a neighbor and stealing eggs from him. They surreptitiously enter the nest, take a mouthful of eggs, swim back to their domain—often with the angry parent in hot pursuit—and deposit the kidnapped eggs into their own nest. This behavior can only be explained in light of females' preference for nests that already contain eggs. The thieving males are trying to make their nest more attractive.[39]

Here is a final example of preference for nests with eggs, but with a twist. In the garibaldi damselfish *Hypsypops rubicundus*, females have this habit of laying their eggs on a flat rock surface that is under the guard of a male. Now, not only do females prefer slates that already support eggs, but they also like to lay next to young eggs rather than old ones. So, let's imagine two males that already have multiple clutches on their slate, with room for more. For one of these males, it so happens that the youngest eggs are on the edge of the slate, separated from the empty space by a row of older eggs. For the other male, the situation is reversed: the old eggs are on the edge, and the empty space is next to the young eggs. This latter male is not bothered because his open crib is next to attractive young eggs. The first male, on the other hand, is in a bit of a bind. His old eggs make the rest of the slate unappealing to females. What to do? Well, as unfatherly as it may be, the garibaldi's answer is: eat the old eggs. Paul Sikkel, who was then at Oregon State University, has documented this behavior. He has reported that male garibaldi that cannibalized such misplaced old eggs ended up with many more new eggs in their nest later, whereas noncannibals remained stuck with a partially empty slate.[40]

Cuckoldry in Fishes

In a great number of species (salmon, sticklebacks, sunfishes, minnows, darters, cichlids, wrasses, parrotfishes, and gobies, among others), some males in the population play the mate-choice game, but others resort to a less flashy alternative. They cuckold the showy males by taking advantage of the fact that in almost all fishes, fertilization is external.[c] During a spawning event between a female and her chosen consort, sperm and eggs are released by the two, but for fertilization to occur, the sperm and eggs must meet in open water. While they are floating in water, however, a female's eggs are vulnerable to interception by the sperm of a cuckolding male.

The cuckolder's tactic is simple. He hides near the periphery of a spawning site, watches what is going on there, and at the critical moment when a male and female shed their gametes, the cuckolder rushes in and releases his sperm. The rightful mate, himself in the act of spawning, is caught unawares. He usually gathers his wits and chases the intruder away, but often the damage is already done: some of the sneaker's sperm have mingled with the female's eggs, resulting in at least some illicit fertilizations.

These "parasitic" males have variously been called sneakers, streakers, cuckolders, hiders, furtive males, accessory males, type II males, or interference-spawning males. Because they are small and superficially resemble females (all the better to fool territorial owners), they have also been dubbed pseudofemales, female mimics, and even transvestite males. They may swim around with authentic females or hide near the nest or territory of the so-called bourgeois males, waiting for their chance to steal fertilizations.

Like group spawners, sneakers have abnormally large testes for their body size. Instead of investing in a large body and defending an attractive territory, they dash in and out and let their sperm do the talking. To maximize their chances of fertilization, they produce a lot of sperm. Despite this, however, they seldom succeed in fertilizing more than 33% of a female's eggs. They are still outcompeted by the bourgeois males, which have the advantage of being next to the spawning female from the very start, and which can therefore better judge the exact moment when she will release her eggs and better synchronize their actions with hers.

Mate choice is an extremely busy field of research in fish ethology, as evidenced by the fact that this chapter is the longest in the book. We have learned that in the eyes of fishes, not all sexual partners are created equal. Fishes have the ability to distinguish between potential mates and to base their choice on criteria that seem to make adaptive sense. They prefer mates that are well built, colorful, symmetrical, healthy, enthusiastic flirts, owners of good real estate, and, in the case of males, already popular with the ladies. This preference leads to a more viable and abundant progeny.

Are you tempted to identify human parallels here, probably helped by the metaphors I have used in the last paragraph? There are indeed similarities, but there is an attending danger in emphasizing them. We may be tempted to believe that just because fishes seem to act like us in some ways, then they must also think and feel like us. We may be tempted to anthropomorphize, that is, to assign human feelings or mental processes to our piscine friends. Doing so is speculative, in my opinion. I am not saying that fishes are pure robots incapable of feelings (love, desire, affection, jealousy, or what-have-you) or even of comprehending the consequences of their choices ("I choose this male because he'll take better care of my young"), but there might also be simple brain mechanisms that could lead to adaptive behavior without necessarily giving rise to heartfelt emotions or conscious rationalizations. The truth is that we do not know what goes on inside the head of a fish. We may very well never know. This chapter is a good place to warn readers of the uncertainty surrounding anthropomorphism because coverage of mate-choice research by the popular press is often rife with expressions borrowed from human affairs—and I am guilty of it too, in this chapter as well as in others within this book. These turns of phrase, although colorful and evocative, may unfortunately create the impression that fishes think and feel just like us. Well, maybe they do to a certain extent, but then again maybe they don't.

10

Shoal Choice

Many animals like to live in groups. Typical examples include large herds of hoofed mammals, flocks of migrating birds, and of course, large schools of fish moving in synchrony. Of these, fishes provide the best opportunity for answering questions about group living. It is relatively easy to keep large groups of fish, manipulate the number of individuals in each group, and expose them to potential predators. Try to do that with sparrows or cows! They just take up too much space. Fish ethologists have therefore contributed greatly to our understanding of group living.

The question that occupies us here is whether it matters to fishes what kind of shoal they are in. Do fishes have the ability to discriminate between shoals and to form aggregations that make adaptive sense?

The first factor in a fish's choice of shoal is shoal size. Bigger shoals should hold many advantages for fishes, the main one being protection against predators. The more eyes there are within the group, the more quickly an approaching predator can be detected. In experiments conducted by Anne Magurran in the United Kingdom, the model of a pike was slowly pulled toward a variety of minnow shoals, and it was observed that the largest shoals ceased to forage earlier during the approach sequence, implying that the pike was detected more quickly. So large shoals do have better early warning systems.[1]

Also, predators are more easily confused when they attack big milling shoals, as indicated by their lower rate of capture success when they initiate contact with larger groups of prey. This has been shown for blue acara cichlids preying on guppies, squid and cuttlefish attacking silverside and mullets, and pike or perch attacking minnows.[2] As long as the shoal remained intact while undertaking evasive action, the predators seemed to concentrate first on one individual and then switch to another, and another, and yet another. These poor predators really gave the impression of being overwhelmed by too much information.

We therefore predict that "smart" fishes should prefer to join larger shoals, especially if there are signs of predators being afield. Mary Hager and Gene Helfman of the University of Georgia in Athens have tested this prediction. They gave fathead minnows a choice between two shoals of various sizes, anywhere between 1 and 28, in many different combinations. A choosing minnow was in a central aquarium sandwiched between two side aquaria in which the shoals were located. Under such conditions, the choosing minnow usually spent more time close to the larger group, especially when the difference in shoal size was pronounced. Hager and Helfman then introduced a fourth aquarium containing a predator (a largemouth bass) behind the middle tank so that the choosing minnow could see it. Under this new condition, the minnow chose the larger shoals more rapidly and spent even less time than before next to the very small groups. The conclusion is that minnows do indeed prefer large shoals, and especially so when they are scared.[3]

Researchers working with redbelly dace and rock bass within a small lake in northern Michigan have obtained similar results but with a twist. By confining prey and predator within jars or enclosures, they could offer single dace a choice between a small and a large group of other dace, with a predatory bass present. The twist was that the predator was always located near the large shoal. In this situation, the dace chose the small shoal away from the bass, but only when the difference in shoal sizes was modest (2 versus 4, or 2 versus 7). If the choice was between 2 and 10, or between 2 and 13, the dace stayed close to the large shoal, even if that meant being nearer the predator. This experiment illustrates how strong the attraction of a large shoal can sometimes be.[4]

The extent to which large shoals are preferred may vary from one species to the next. Would we expect the choice of a creek chub, a vulnerable minnow that likes to shoal, to be the same as that of a three-spined stickleback, a fish that is well armored with external bony plates and dorsal spines to confront predators and that shoals only occasionally? It turns out that both species, after being scared in the laboratory by the simple trick of turning off the lights for half a second, swim toward the larger of two shoals. There are, however, subtle differences. Sticklebacks cease to show a preference for large groups if the two shoal sizes differ by very little (5 versus 6, or 5 versus 7, rather that 5 versus 9, or 5 versus 10). They also have trouble if little time is made available for them to examine the two shoals before being forced to choose (10 or 20 seconds instead of 150 seconds). Chub, on the other hand, can maintain their preference for the larger shoal even under such stringent conditions. It may be that chub derive greater antipredator benefits from shoaling than do sticklebacks, and this translates into a more finely tuned shoal-choice behavior when predation risk raises its ugly head.[5]

Bring Them On! Predators That Are Not Deterred by Big Shoals

Some predators have developed strategies specifically designed to counter the antipredator advantages of big fish shoals.

They beat the confusion effect by forgetting about individuals, relying instead on rapid strikes more or less at random in the thickest part of a shoal. For example, marlin and swordfish cut a swath through shoals, stunning prey with systematic swings of their saw or sword. Thresher sharks do the same with their tails. After the rest of the shoal has fled, these so-called impact predators return and pick up the prey they have stunned (hopefully, other predators will not have parasitized their effort and snatched the lifeless prey already). Dolphins, seals, whales, and pelicans are also known to actively herd fish shoals so that they are able to randomly bite into the shoal's compact mass.

Fish-eating birds that dive from high above may also like big shoals because they are probably easier to spot from the air. Prey fishes may feel no safer in a bigger group when faced with the threat of aerial attack. In one experiment on guppies, fish that were exposed to the passing shadow of a kingfisher seemed to be just as scared when in a group as when alone, as manifested by the similar duration of their freezing time and the depth at which they took station.[a]

So-called ambush predators can deal equally well with big and small shoals. One laboratory study showed that predatory rock bass did not care whether they faced groups of 2, 3, 4, or up to 13 creek chubs. As long as they could launch their attack from behind a screen of black plastic folds, which meant at close range to the unsuspecting shoal, the success rate of the rock bass was the same irrespective of prey shoal size.[b]

One way for predators to minimize confusion when they attack a shoal is to concentrate on stragglers that become separated from the shoal, as confirmed by numerous field and laboratory observations.[6] They can also focus on individuals that look different from the rest, called the oddity effect. Largemouth bass, for example, are known to preferentially attack minnows that are differently sized or differently colored within a given shoal.[7] To counter this oddity effect, wary prey should try to hang around with companions that look the same as they do. Literally, fishes do not want to stand out in a crowd.

Esa Ranta and co-workers at the University of Helsinki have tested this shoal-choice hypothesis with brook, ten-spined, and three-spined sticklebacks. When given a choice between large and small conspecifics, small sticklebacks spent more time close to the other small fish, especially when a predator (a rainbow trout) was shown to them. Large sticklebacks also preferred similarly sized conspecifics, although this preference was not always enhanced in the presence of the predator.[8]

The Finnish researchers followed up their experiments with an analysis of the size distribution of wild shoal members. With a sweep net, they captured 24 shoals of juvenile three-spined sticklebacks from shallow areas in the Baltic Sea and found that members of each shoal were more similar to each other in body size than could be expected by chance. In other words, instead of finding shoals that comprised both large and small fish in more or less even numbers, they caught shoals that featured almost exclusively small or exclusively large individuals. Such a result is consistent with the idea that fish try to congregate with other individuals of similar size.[9]

This preference for similar conspecifics is especially obvious in multispecies shoals. Various species of minnows can become mingled in the course of their normal activities, but if a realistic-looking pike model is dragged through the water, the fish suddenly reorganize themselves and form big groups in which each individual ends up being closer to members of its own species. This behavior was observed by John Allan and Tony Pitcher from the University of Wales at Bangor, who used videotapes to accurately measure distances between individuals.[10] (On videotapes, distances are often measured in units of fish body length, as seen on the television screen, and therefore it helps to have minnows of uniform size. In addition to the frontal view, a top view is also provided by the use of mirrors placed at a 45° angle above the tank, allowing analysis in three dimensions.)

The previous example of species segregation involved minnows of similar size. Jens Krause, who has conducted numerous studies on the shoaling behavior of fishes in Germany, England, and Canada, wondered what would happen if a fish had to choose between individuals of the same species but of a different size versus individuals from a different species but of a similar size. Which criterion would prevail, species similarity or size similarity? In the laboratory of Jean-Guy Godin in Canada, Krause offered banded killifish a choice between four other killifish larger than the choosing fish and four golden shiners the same size as the choosing fish. Visually, golden shiners cannot be confused with killifish. To be properly motivated, the choosing fish were exposed to the model of a heron head, and their subsequent shoal choice was measured. The killies ended up spending more time close to the similarly sized shiners, expressing the overriding influence of size similarity over species similarity.

On a roll, Krause and Godin asked whether size similarity could even be more important than shoal size. They already knew that their killies preferred shoals of eight to shoals of four conspecifics. But what would happen if the size of the fish in the small shoals matched the size of the choosing fish more closely? The killies ended up preferring such small shoals. Therefore, size similarity is so important that it takes precedence over shoal size, at least in this species and at those shoal sizes.[11]

In nature, fishes probably try to satisfy all three criteria—large numbers, species similarity, and size similarity—in their group formation. With help from David Brown, Krause and Godin analyzed shoal composition in the lake from which their experimental fishes came. They captured no less than 34 shoals by encircling them, one group at a time, with a beach seine net. Approximately 80% of these shoals had more than 10 members—one mammoth group tipped the scale at 776. These shoals turned out to be more homogeneous than would be expected by chance, in terms of both species identity and their members' body size. Although five species were present in the lake, most shoals were comprised of only one or two species. Even within the two-species shoals, all fish tended to be of the same size—either all small, for example, or all large.[12]

Another possible advantage to group living is the presence of more food-finders that can share—if only involuntarily—their discoveries. If food is hidden in the gravel of an aquarium and shoals are let loose to search for it, the food is always found and dug up earlier by the bigger shoals. And almost invariably, as soon as the food is discovered by one individual, fish from all around quit their own search and join the discoverer, crashing its party.[13] Therefore, hungry fish might benefit from belonging to larger groups. However, there is another side to this coin: if there are too many fish in a shoal, competition for food may become too severe. So, although big shoals are advantageous to start with, competition may actually set an upper limit to the ideal group size.

If the shoal members are particularly hungry, they may deem the level of competition found in large shoals to be unacceptable and so may prefer smaller groups. Indeed, when left on their own, large numbers of hungry fish usually break up into shoals that are smaller, less compact, and less coordinated, because each fish tends to strike out more on its own, looking for food that it can keep to itself.[14] They may then become more vulnerable to predation, but—damn the torpedoes—they have to eat!

In fact, fish in a shoal seem to be acutely aware of exactly how much competition is going on around them relative to the amount of resources available. This can give rise to a phenomenon called the ideal free distribution; that is, if more than one patch of food is available in the environment and these patches differ in their probability of reward, then groups of fish should split up and distribute themselves over the various patches accordingly. Thus, compared with elsewhere in the environment, twice as many fish should be found in a patch that yields twice as much food.

It is relatively straightforward to test the ideal free distribution hypothesis in the laboratory. Tubes can be installed at both ends of an aquarium and eggs (from insects or other fish) can be dropped through these tubes at different intervals of time. The shorter the interval, the richer that feeding station is. Shoals of 6–10 fish are then introduced into the tank and allowed some time to sample the two stations. Thereafter, all that is left to do is to count the number of fish at each end. If one station dispenses twice as many eggs in the same period of time as the other, then out of a group of 6 fish, for example, 4 should take position at that station and 2 at the other one (see fig. 10.1).

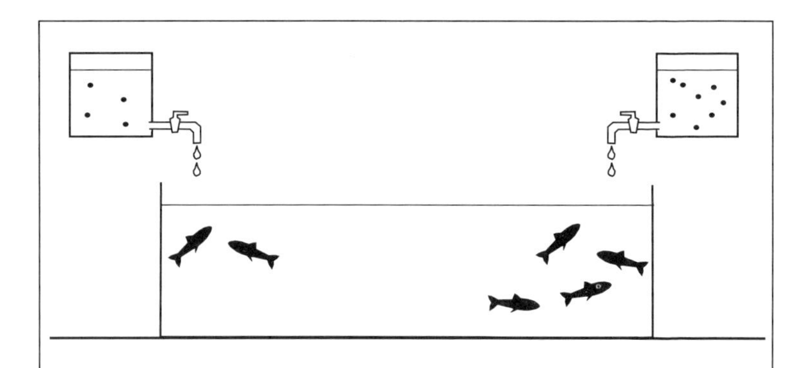

Fig. 10.1. Fishes can evaluate the richness of a food patch and the degree of competition they have to face there. The result is that nonaggressive individuals often distribute themselves between food patches in accordance with the profitability of each patch. If twice as much food is made available at one end of an aquarium, twice as many fish will gather there.

Predictions of this kind have been verified in sticklebacks, guppies, and goldfish. The phenomenon is also present in cichlids and salmon, although in their case the results are muddled by the influence of dominant individuals, which tends to exclude other fish from the rich stations. This exclusion effect can be abated through the use of multiholed tubes that dispense food at unpredictable places over a large area in the food patch, making the food more difficult to monopolize.[15]

So what is the ideal group size for a hungry fish? When does the disadvantage of increased food competition balance out the advantage of increased food discovery as shoal size increases? The answer obviously will vary from species to species, but with sticklebacks we have some idea. At Laval University in Quebec, Nathalie Van Havre and Gerry FitzGerald offered female three-spined sticklebacks a choice between shoals of 15 or 45 conspecifics. They observed that satiated females preferred the large group (good antipredation choice), whereas females unfed for 24 hours were more attracted to the smaller group (good anticompetition choice). Meanwhile, Jens Krause obtained different but not necessarily contradictory results: in test choices of 5 versus 3, 10 versus 3, or 20 versus 3 individuals, both satiated and hungry sticklebacks preferred the larger group, although the preference was less marked in the case of the hungry fish. When we take the two studies together, we are left with the picture that compared with a shoal size of 3 or 45 fish, a shoal size of 10–20 fish may represent the ideal compromise for a hungry stickleback that wants both to increase the chance of finding food and to decrease competition.[16]

Cooperation and Competition in Groups of Hunters

In the laboratory, bluegill sunfish are reported to capture more amphipods per fish as their group size increases from 1 to 4. Gary Mittelbach, the author of that report, attributed the result to a flushing effect. The prey were hiding in patches of vegetation set up in a large aquarium, and they were more easily scared out of their hiding places by the combined activity of many fish. However, the per capita capture rate nose-dived when an even larger group of 6 sunfish was tested, because the fish then started to quarrel among themselves. In sunfish, the optimum group size for cooperative hunting may be 4.[c]

On coral reefs, damselfishes zealously defend individual territories that contain mats of algae and sometimes also their own nest with eggs. These relatively small fishes are so fearless that they may attack divers. A single damselfish has no trouble repelling lone parrotfishes, surgeonfishes, and wrasses, but these latter species sometimes cooperate and form groups of 30–300 individuals that can overwhelm the defenses of the damselfish and devour its algal garden or its eggs. In many cases, these large groups do not form when damselfishes and their resources are absent or less abundant, which suggests that the groups are formed with the specific intent of swamping the valiant defenders.[d]

Shoaling fishes in the examples above may have been cooperating, yet they were not necessarily coordinated. But there is at least one example of a marine hunter that displays an amazing degree of coordinated action while pursuing prey. This predator, the amberjack *Seriola lalandei,* works in packs, actively trying to split up prey schools to create smaller pockets of confusion. Packs of 5–10 amberjacks form U-shaped lines that cut off the tail end of prey schools and herd the unfortunate stragglers next to seawalls. There the amberjacks proceed to attack their prey in a disciplined manner, taking their turn to lunge at individual prey in the center of the newly downsized group.[e]

Competition is a function not only of shoal size but also of the body size of its members. Within the same species, larger fish outcompete smaller ones. So we might expect hungry fish to join groups of smaller conspecifics and try to steal their food. But we have already seen that fish do not like to be with other fish of a different size because then they become more conspicuous to predators. Should a hungry fish be willing to risk the oddity effect by joining smaller partners to decrease food competition?

Nancy Saulnier and I have tackled this question through some work with golden shiners. When offered a choice between a group of five small individuals and a group of five large ones, large shiners, if they were well fed, preferred the other large fish; but they reversed this preference and chose to be with the small ones after being deprived of food for 48 hours. Small test fish preferred other small fish, no matter how satiated or hungry they were (in their case, both the oddity effect and food competition dictated that they stick to similarly sized shoalmates). We concluded that hungry fish are more willing to risk the oddity effect in their shoaling behavior, as long as they can gain a competitive edge by doing so. If a shoal splits up in nature, a large fish may quickly assess the relative body size of members of each subgroup and stay with the appropriate one according to its hunger level.[17]

The behavior of potential shoalmates may also affect a fish's decision to join the group. When given a choice between a shoal that is having a feast versus one that is idle or feeding less, fish try to mix with the actively feeding one.[18] Admittedly, such a choice is not very surprising, but the discriminating powers of fish are in fact more refined than that. Let's take the case of food-anticipatory activity. When fish are fed at the same hour every day, they begin, after a few days, to anticipate food arrival, becoming more active and more oriented toward the food source a few hours before mealtime—as we saw in chapter 6 on telling time. Bruno Gallant and I have shown that this subtle behavior can be used as a cue by test fish when they choose a shoal. In the morning, we offered golden shiners a choice between a shoal of fish that was accustomed to being fed in the morning versus one that was always fed in the afternoon. The tests were repeated in the afternoon. Food-anticipatory activity was present: the shoal that was used to being fed at a particular time of day turned out to be more mobile at that time and to spend more time in the upper half of the water column than did the other shoal (food was normally delivered at the surface by automatic feeders, although it was not given on test days). During morning tests, the choosing fish ended up spending more time close to the morning-fed shoal, and in the afternoon they preferred the afternoon-fed shoal. The hungrier the test fish was (test fish were deprived of food for 48, 24, or 1 hour), the more pronounced this preference was.[19]

We can therefore envision natural scenarios in which one shoal meets up with another and decides whether or not to combine with it or not based on the first shoal's hunger level and the second shoal's behavior, be it foraging behavior or food-anticipatory activity. A hungry shoal would merge with a food-anticipating shoal, whereas a satiated group might continue on its merry way.

Fishes may also stay away from certain shoals based on the shoal's behavior. For example, parasitized sticklebacks often behave abnormally, and they are shunned in shoal-choice tests. External signs that betray parasitic infection may also curtail the attractiveness of shoalmates. When Jens Krause took banded killifish and injected a speck of black ink underneath each one's skin, simulating the mark of a parasite, these killifish still displayed the normal behavior of healthy fish but were nevertheless rejected by their shoalmates in choice tests.[20]

Follow the Leader

The ability of gregarious fishes to use the behavior of shoalmates as a cue that food has been or is about to be discovered gives rise to the possibility that some individuals that seem to know what they are doing might act as leaders in a group. I have been able to demonstrate this in the following manner. In a large basin I placed 12 golden shiners. Having never been in this basin before, the fish were nervous and spent the whole day in a corner that was in the shade (a common reaction in wary fish). In the brightly lit corner opposite, I set up a feeder that delivered food flakes at noon. After many days, the fish learned to leave the shady corner around noontime and seek food in the lit corner. They did this even on test days when food was withheld (an example of their internal clock at work). Then I replaced 11 of the 12 fish with naive individuals that had never been in the basin before. You might expect these newcomers to stay in the shade all day, but instead they followed the remaining informed individual to the food corner around noontime, even though no food had been delivered on that day. It seems the old hand had acted as a leader for the rest of the group.[f]

Parental male sticklebacks make good use of the parasitic tendency of fish to share the spoils of a food discoverer. In sticklebacks, males assume the duty of taking care of eggs inside a nest that consists of plant material lying flat on the ground. In some populations, roaming shoals of hungry females sometimes fall upon the nest of a parental male, thoroughly devastating it and eating all the eggs inside. Males take a dim view of this behavior, and they have worked out a defensive ruse. When a male sees a menacing shoal of hungry-looking females coming his way, he swims a short distance away from his nest and starts poking his snout into the ground. This is the same action a female would perform while raiding a nest. The display commonly fools the females into believing that a nest has been discovered. They rush to that site and start digging there too. Meanwhile, the male leaves the writhing mass of females behind and returns to his territory, hoping—consciously or not—that the cloud of sediments lifted by the feeding frenzy will conceal his own very real nest. This striking behavior is similar to the broken-wing display used by ground-nesting birds to lure predators away from their nest.[g]

In any shoal, front and back positions are readily told apart, and to a certain extent, so are central and peripheral ones. The possibility arises that different positions may confer different advantages. Can fishes discriminate between within-shoal positions? Here the main hypothesis underlying research has been that fish in the periphery are more likely to be the first ones to encounter food but also the first ones to be attacked by predators.

To provide support for this idea, Jens Krause took advantage of the presence of a mixed shoal of roach and dace that consistently fed in the fast-running water of a small stream near Cambridge in the United Kingdom. The water was clear, and by simply sitting motionless on the bank, Krause could observe the activity of the fish. He saw that individuals at the front picked up food particles more often, especially drifting food, as we might have expected. Krause then captured 28 roach from the shoal. Half of these were marked with a blue spot near the tail (by a subcutaneous injection of Alcian Blue) and kept unfed in the laboratory for 3 days. The other half were identified with a blue spot near the head and fed in abundance on commercial dry food for 3 days. All were released back into the stream. Two-thirds of them took off and were never seen again, but the rest (7 starved ones and 6 well fed) stayed in the shoal. On the first day, the starved ones were seen to spend most of their time at the front, whereas the well-fed ones formed the rear guard. Three days later, after the feeding status of all fish had presumably equilibrated, there was no more distinction in shoal position.[21]

This experiment convincingly showed that front positions are perceived by fish as more advantageous for obtaining food, at least for shoals subsisting mostly on drifting food. With his colleagues Dirk Bumann and Dietmar Todt, Krause also showed that even in the stagnant waters of an aquarium, hungry roach tend to be at the front of a shoal and to lead its peregrinations, and to enjoy higher feeding rates (on *Daphnia*) in the process.[22]

Why would well-fed fish stay away from the front of a shoal? Perhaps it is easier to swim or to fight currents in the wake of other fish, but this is still a controversial and unresolved issue. Maybe from the back they can more easily keep an eye on their shoalmates and thus witness their fright reaction to a predator. And maybe the leaders of a shoal are more likely to be targeted by ambush predators. Bumann, Krause, and Dan Rubenstein have let groups of creek chub swim in an aquarium that also contained a predatory rock bass lurking in folds of black plastic. In the 30 trials that were conducted (with a total of 10 different bass acting as predator), the ambush predator always attacked an individual chub from the front half of the shoal. In fact, for 25 of those 30 trials, the target was the leader at the very head of the shoal.[23]

Therefore, we can see why a well-fed fish might be willing to let other individuals take the lead. Hungry fish may accept the risk of leading in return for a higher probability of food discovery, but a well-fed fish feels no such compulsion. However, although frontal attacks may be the typical behavior of an ambush predator, more active hunters are known to approach shoals from the rear, so the relative safety of front and back positions probably varies from one ecological situation to the next.

Positions in the very center of a shoal may give the most protection. At least they seem to be preferred by frightened individuals. In an ingenuous experiment to prove this, Krause made good use of the scaring potential of alarm pheromones in dace (see chapter 1 on olfaction). He also capitalized on the fact that fish can habituate to this chemical signal: they stop reacting to it after having repeatedly smelled it without any further sign of predation. Krause kept one minnow that was not habituated to the alarm pheromone with 14 dace that were. He then pumped a solution of minnow alarm substance into their tank. The habituated dace did not significantly alter their behavior, as expected, but the naive minnow did: it quickly moved to a central position among the rest of the shoal (see fig. 10.2). If minnows shift to the center of a shoal when they are scared, it must be that they see such places as safer.[24]

A less common form of aggregation is the colonial nesting of some fishes. The main advantage of large breeding colonies seems to be protection against predators through cumulative nest defense. On reefs off Oahu, larger colonies of the Hawaiian sergeant, also known

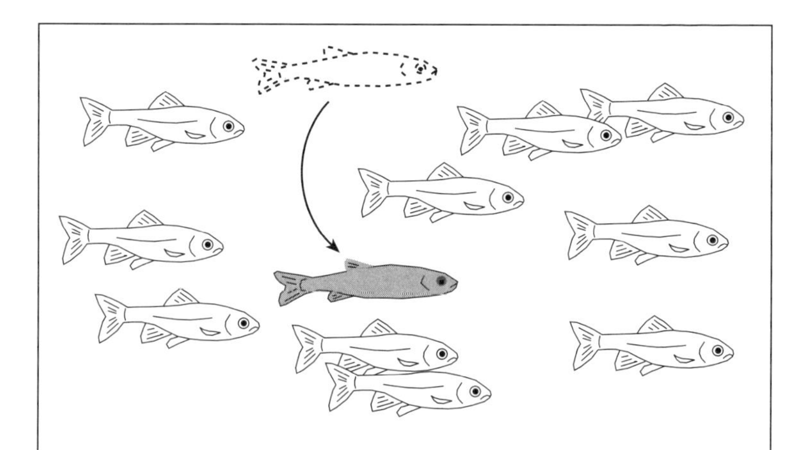

Fig. 10.2. In shoaling fishes, wary individuals often prefer central positions. If alarm substance is poured into an aquarium, the fish that have already been habituated to this signal will not alter their general position, but others, such as this lone European minnow in a shoal of habituated dace, will move closer to the spatial center of the group.

as the maomao, usually suffer fewer intrusions by egg predators. If some parental male sergeants are removed from various colonies and their eggs are left behind, these eggs survive longer if they happen to lie within bigger colonies. The same applies to nesting colonies of bluegill sunfish. The eggs survive longer because in large colonies there are more males available to ward off predators that approach the site.[25]

As with shoals, we might expect a greater rate of predatory attacks on the periphery of the colony, both because this is the first encounter point between egg predators and colony members, and because peripherals have fewer neighbors to help them repel attackers. Studies of both Pacific sergeant majors and bluegill sunfish have confirmed that higher predation rates befall peripheral rather than central nests. This, by the way, should create selective pressure for females to mate preferentially with males that own centrally located nests, and this is in fact what females do.[26]

We have seen that there are a number of criteria used by fishes for adaptive shoal choice: shoal size, species or body-size similarity, shoalmate health, and shoal behavior. The influence of many of these factors can be modulated by the hunger level of the fish making the choice—for example, hungry fish may not prefer large shoals as much as well-fed individuals do, even though the smaller shoals in which they end up may render them more vulnerable to predators. In this we see an example of a notion that is more thoroughly explored in the next chapter: consciously or not, fishes are capable of making compromises, depending on their individual status or that of their environment.

11

Making Compromises

The life of fishes is dictated by a triumvirate of imperatives: the need to reproduce, the need to eat, and the need to avoid being eaten. Life would be rosy if fishes could fulfill all of those three obligations simultaneously and indulge in whatever they wanted to do, whenever they wanted to do it with careless abandon. But of course, life isn't so simple. Some activities are not compatible with each other or cannot be done at the same time. So fishes are faced with choices in the organization of their day planner. And sometimes trade-offs must be accepted when deciding what to do at any given time. We do not know how fish brains "calculate" and "weigh" the pros and cons involved in any difficult decisions they make, or what rules they may follow to solve problems, but at least we know that fishes are capable of well-adapted compromises even if they may not be "aware" of the consequences of their decisions.

The most common compromises that fishes must make are related to the annoying influence of predation risk. The big problem here is that animals of all sorts find fish delicious. Herons, kingfishers, mergansers, and all marine birds like to eat fish. Marine mammals, minks, and bears do it too. Big fishes themselves are piscivorous. As a taxonomic group, fishes face one of the most diverse array of predators imaginable. Threat comes from below and from above, during the day and at night, and at almost all stages of life. Is it any wonder that most fishes—certainly all the small ones—are skittish creatures? When placed in a new environment, most wild fishes cower in nooks and crannies and don't dare raise a fin. They do not want to draw the attention of predators.

This puts a cramp in their lifestyle. First of all, they cannot always eat where they want. This has been shown by several studies that have looked at the distribution of fishes in ponds, both in the presence and absence of predators. These studies were conducted with bluegill sunfish (with largemouth bass as the predator), minnows (with pike), young crucian carp (with Eurasian perch), and young Eurasian perch (with their cannibalistic elders). In all cases, prey in the predator-free condition blissfully occupied open waters as well as shallow weeded areas, whereas prey exposed to predators remained exclusively in the shallow weeded areas because that is where they were best protected. Confinement to the shallows usually led to slower growth rates because food was not as plentiful there and because there was more competition for it from the great concentration of refugees.[1]

In the above experiments, some prey could be seen to venture, sometimes even to take up residence, in the open waters in which the predators operated most efficiently. Invariably these fearless individuals were large. Great size confers some degree of immunity against predators. These large individuals were not constrained by the competitive bottleneck that affected their smaller brethren in the shallows. Therefore they grew more quickly and became even safer from danger, a case of the rich getting richer and the poor staying poor.

The antipredator benefits of large size are further illustrated by an intriguing observation. In the absence of predators, growing crucian carp develop slim bodies that are hydrodynamically efficient. But in the presence of predators, carp grow to become rounder and larger, a body shape that is not so good for swimming but more likely to deter predators because it is not so easy to swallow. It seems the differential growth that leads to rotundness is induced by exposure to skin substances, possibly alarm pheromones from other carp, exuding from the feces of the predator.[2]

Predation experiments can also be performed in the laboratory. Within wading pools it is possible to construct various patches of weeds by using green polypropylene ropes attached to grids of wire mesh. Densities of 50, 100, 250, 500, or 1000 stems per square meter can be achieved. Then one need only examine the distribution of a prey species—bluegill sunfish, for example—before and after the introduction of a predatory largemouth bass into the pools. Vytenis Gotceitas did this during his Ph.D. studies at Queen's University in Ontario. He observed that his bluegills normally patrolled in low-density weed patches—this is where they had the most success finding and catching the damselfly nymphs that Gotceitas had thoughtfully provided. But when the predator was thrust upon the scene, most bluegills moved to the high-density weed patches, even though they ended up suffering a poorer foraging rate there.[3]

If habitat switches reflect a trade-off between foraging and avoiding predation, then it should be possible to set up an experiment that manipulates this balance and tip it either in favor of more foraging despite the risk of predation or, conversely, in favor of more sheltering despite the risk of starvation. The simplest way to do this is to compare the behavior of hungry and satiated fish. Both can be offered a choice between spending some time in a safe habitat devoid of food or in a risky one in which there is food. This experiment has been conducted in the laboratory, with crucian carp facing pike, black gobies facing cod, pink salmon fry facing adult chinook, and juvenile coho salmon facing adult rainbow trout. In all cases, the hungry individuals spent more time in the risky area, close to the predator but with good access to food, than did the better-fed fish.[4]

Those experiments hint at another way to affect the balance of foraging opportunity and predation risk. We can vary the quantity or the quality of food in the risky habitat. A choice can be offered between two patches, one that gives access to a little food and is placed near an adjacent aquarium that contains no, or maybe only one, predator, versus another patch that offers more food but is also next to an aquarium containing two predators. The question is: How much more food should the dangerous patch contain to draw the wary prey there?

Experiments of this kind have been performed with juvenile creek chub facing predatory adults, young black surfperch at risk from kelp bass, European minnows exposed to a kingfisher, guppies facing cichlids, and upland bullies viewing a salmon. The prey's switch from safe to dangerous habitat took place only when food was at least three- to fourfold, and sometimes as much as 28-fold, more abundant in the risky site, a substantial difference.[5] This could explain why, in the natural experiments described above, fishes at risk from predation stayed in the shallow areas of ponds and lakes despite the lower food supply there. If fishes have at least enough food to survive in the safe habitat, and the dangerous habitat is not that much better in terms of food availability, then prey may elect to stay in the safe habitat most of the time.

Another way to tip the scale is to alter the availability of refuges in the various habitats. Prey may accept venturing into predator-rich areas if there is also structure there to protect them. Douglas Fraser and Richard Cerri built compartmentalized channels within a spring-fed stream in the Hudson–Mohawk River watershed. Within each compartment they could manipulate the presence or absence of a predator (adult creek chub) and the structural complexity of the habitat (pieces of black pipe, wood, and covers providing shade). The compartments were separated by wood dividers with slots big enough to allow small minnows to go in and out but too small to allow the predators to exit. Small minnows (young creek chubs and blacknose dace) were let loose in those channels, and they were free to move from compartment to compartment. Their distribution could be determined at any time by dropping hinged gates that effectively made all fish prisoners of the chamber in which they happened to be at that moment. In this way, Fraser and Cerri observed that minnows tended to avoid compartments with predators but that this avoidance was less marked when structure was present. Predator avoidance is a strong incentive at all times, but the presence of potential hiding places can mitigate it somewhat. Similar results have been obtained in the laboratory with other species.[6]

Habitat shifts are a good example of compromise, but unfortunately they may not afford complete safety. Some predators have a nasty habit of adapting and venturing into the areas in which their prey take refuge—predators have to make a living too, you know.[7] For example, largemouth bass can switch from cruising in open waters to ambushing in vegetated areas. Small prey fish may flee from harmful perch in open waters only to fall prey to a stalking pike in the weeds. Minnows may think they are safe from large predatory fishes in the shallows, but then they are nabbed by a heron. As I said earlier, fishes are just too tasty. They are almost never completely safe. Nevertheless, the fact remains that habitat switches are a compromise that can at least contribute to lower predation risk. Better a small risk of being caught by a pike in the weeds than guaranteed death from a perch in open waters.

In response to predation risk, some fishes may find compromises other than habitat switches to be more palatable. For example, they may decide to stay where they are but to become less conspicuous, even if that means feeding or courting less effectively.[8] One example comes from coho salmon. In this species, juveniles (parr) usually hold station somewhere in a stream and occasionally dash upstream to intercept drifting prey. Researchers at Simon Fraser University in Vancouver compared the foraging behavior of young cohos that could feed under two different conditions, either undisturbed or after being distracted by the presentation of a photograph depicting an adult rainbow trout (a predator of young salmon). The results were that, all other things being equal, the cohos that had seen the photograph were not willing to swim as far away as usual to catch drifting insects (see fig. 11.1). Whereas unperturbed salmon were willing to swim 25 cm upstream on average to catch a big fly, disturbed salmon would only go 16 cm.[9]

Here again it is possible to alter the trade-off between safety and foraging by playing with the hunger level of the salmon or with the levels of predation risk. The Simon Fraser University researchers, Larry Dill and Alex Fraser, manipulated their cohos in this way. Hungry salmon ended up reducing their attack distance on drifting prey when scared by a predator, as expected, but not as much as better-fed individuals did. Because the cohos were hungry, they were willing to take a little bit more risk. Salmon that could see their own image in a mirror were also willing to take more risk by dashing a little farther than lone individuals. Either they perceived the mirror image as a competitor for food and consequently were more motivated to get the food,[10] or they felt safer because they had a companion and reckoned there was less chance of them being the specific target of an attack.

The scientists also manipulated the balance in another way: they varied the frequency with which the predator image was presented. As expected, salmon that were exposed to the image of a predator more often (every 22 minutes) reduced their attack distance to a greater degree than did salmon that saw the predator less frequently (only at 45-minute intervals). The fish were able to estimate the higher level of risk and adjust their foraging behavior accordingly.[11]

The need to avoid conspicuous behavior in the presence of a predator can have an impact on a fish's sex life, or more precisely, on its courtship behavior. Let's take the case of guppies. Males have two ways of mating with females. They can woo them with a sigmoid display, in which the body is arched and the fins are extended. Such a display is conspicuous, can take up to 5 seconds to perform, and must be done fairly often before a female finally agrees to mate. The second strategy

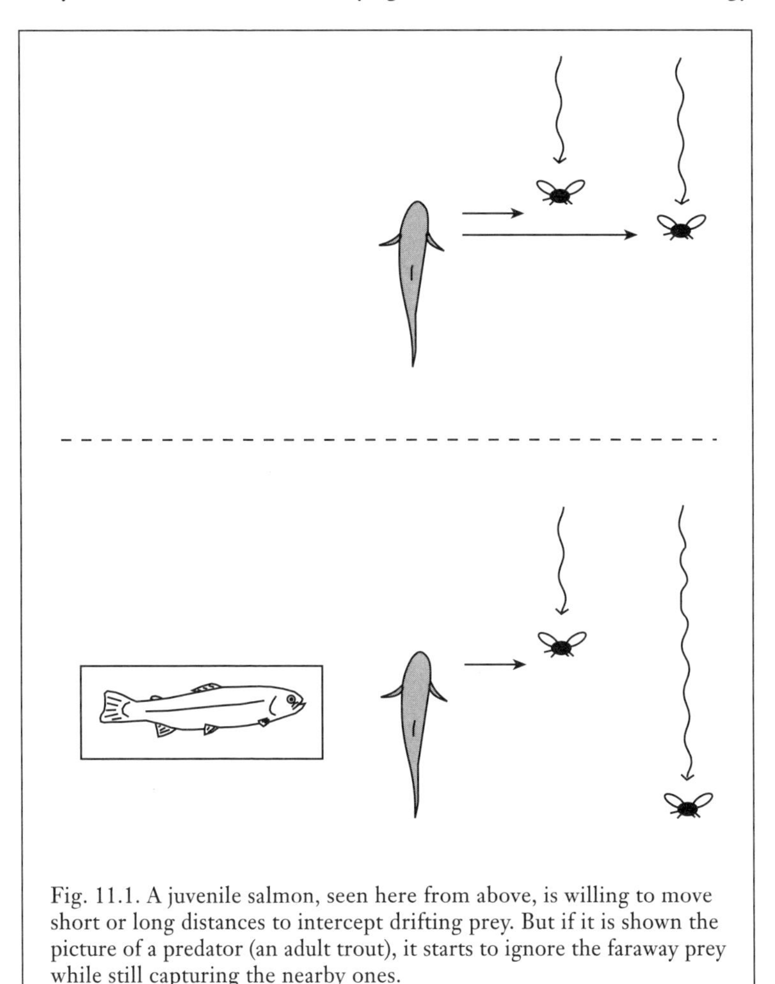

Fig. 11.1. A juvenile salmon, seen here from above, is willing to move short or long distances to intercept drifting prey. But if it is shown the picture of a predator (an adult trout), it starts to ignore the faraway prey while still capturing the nearby ones.

is sneakier. It is called gonopodial thrusting, a forceful insemination without the female's cooperation. Gonopodial thrusting is less conspicuous than sigmoid displays, but the chance of a successful insemination is also reduced because the female tries to resist it. The interesting point here is that when we compare the relative frequency of both strategies in the presence and in the absence of predators (cichlids or characids) at large in the same environment, the sneaky behavior predominates when predators are present, whereas the conspicuous display is more important in the predator's absence.[12]

It seems therefore that wary fish abandon effective but conspicuous courtship displays and resort to less showy but safer alternatives if they can. Another option when fish are under predation threat is to shorten the duration of courtship before finally mating, as has been observed in the pipefish *Syngnathus typhle*, in sand gobies, and in sticklebacks.[13]

There is one category of fishes for which reduced activity is an integral part of antipredator strategy: cryptic species whose body color matches the surroundings. For camouflage to be effective against a static background, the fish must itself remain motionless. There is evidence that freezing in cryptic species is a true effort to blend in and not simply an attempt to reduce conspicuous movements irrespective of the potential for camouflage. Tidepool sculpins, whose body markings mimic the appearance of sand, have been kept in aquaria with either a matching (sandy) or nonmatching (white) bottom. When scared by the introduction of an alarm substance, the fish on matching sand reduced their movements as might be expected. However, the fish on the nonmatching white substrate did not alter their activity. For them, immobility would have conferred no cryptic advantage, and consequently that tactic was not adopted. Active search for a refuge was a better alternative in that case.[14]

Another example is provided by three darter species. The fantail, greenside, and orangethroat darters wear dull colors outside of the breeding season, and all three species freeze over mucky bottoms in response to predator signs. During the breeding season, the male fantail and greenside darters develop a conspicuous green body color, but because they normally breed near matching green algae, they continue to freeze when alarmed. In contrast, male orangethroat darters develop intense orange, blue, yellow, and red breeding colors. Needless to say, they cannot find matching surroundings, and therefore it comes as no surprise that they abandon freezing as an antipredator tactic and resort to fleeing instead.[15]

Sometimes, fishes are caught between a rock and a hard place. For example, when oxygen levels become too low, many fishes resort to air breathing or to what is known as aquatic surface respiration (that is, breathing very near the surface, where water holds more oxygen because of its contact with air). Proximity to the surface means predation risk from aerial and terrestrial piscivores, and accordingly fishes resign themselves to approaching the surface only when dissolved oxygen levels get really low, below 25% of their normal values. And even then, they can moderate the frequency of breathing near the surface if they perceive the presence of a fish-eating bird nearby. To demonstrate this effect of predation, Don Kramer and his students at McGill University in Montreal hand-reared a green heron until it was one year old and allowed it to forage in pools stocked with various kinds of fishes (10 species of tropical fishes as well as bluegill sunfish and central mudminnows). The researchers experimentally reduced dissolved oxygen to either 1.6 or 0.5 parts per million (ppm) (or mg of oxygen per liter of water; a normal level would be approximately 8 ppm). As expected, the fishes came to the surface, but not as often as they did when the heron was not there—the fishes were aware of the bird's presence and the danger it represented. Yet they were still forced to come up once in a while to perform aquatic surface respiration or air breathing, especially when oxygen in the water was at 0.5 ppm. The heron caught more fishes at 0.5 ppm than at 1.6 ppm because of the prey's greater use of the surface at that concentration.[16]

With this compromise between predation risk and air breathing in mind, it is worth mentioning an observation by John Gee of the University of Manitoba. He has documented that mudminnows, when in a group, start to synchronize their air breaths after being disturbed by a simulated heron strike (the wooden model of a heron head was plunged into the water). Instead of each individual breaking the surface at any time, the mudminnows scared by Gee's fake heron tended to break the surface more simultaneously. These wary mudminnows may have perceived the risk of predation and decided to take a gulp of air only after seeing another fish get away with it.[17]

One type of behavior that offers rich possibilities for the study of compromise making is parental care. Caring for the eggs or the young—cleaning them, fanning them, defending them against potential predators—is a parental endeavor that is costly in terms of both energy and risk to life. Vigilance and defense must be constant because eggs and young represent nutritious little snacks in the eyes of other fishes, which try hard to get at them. Because of the cost of their care, parents must be judicious about how much parental effort they award their progeny.

In particular, it might be disadvantageous for a parent to devote too much care to a current brood at the expense of the parent's potential for future reproduction. Put more bluntly: if a parent were stuck with a small brood, it might do better to limit the energy invested in the care of such a low-yield evolutionary prospect and instead save itself for better attempts in the future. This and other related ideas come under the banner of parental investment theory. In the late 1980s and early 1990s, the scientific literature saw a burst of publication on this topic. Many of the published articles supported the notion that parents could indeed adjust their level of care as a function of the "value" of their brood.

Many studies—at least one each on a greenling, a damselfish, a darter, a minnow, two cichlids, and a bully—have shown that parents with small egg clutches under their care are more likely to cannibalize these eggs, eating all of them as if to prepare for a more successful attempt as soon as possible.[18] In the laboratory, the egg clutches of convict cichlids can be artificially diminished or augmented by transferring eggs from nest to nest. When the resin-coated model of a predator is lowered into the tank (or, in the case of one study, moved in the water by a toy car running on a portable track resting on top of the aquaria), the parents with reduced clutches are seen to direct fewer aggressive acts toward that predator, whereas those with augmented clutches attack the predator more ferociously.[19]

Compromises can be expressed early in the reproductive cycle. Witness the work of Brian Wisenden during his graduate studies at the University of Western Ontario. Wisenden was studying convict cichlids. He forced his convicts to lay eggs either in a secure spawning "cave" (an overturned flowerpot with only one small triangular opening at the rim) or in a risky one (an overturned flowerpot again, but with two large openings). The risk stemmed from an inability to defend both openings simultaneously against egg predators. Wisenden counted the number of eggs laid in each type of cave by various females, and he found that fewer eggs were entrusted to the protection of risky caves (see fig. 11.2). It was as if the females knew that the nest was less secure, sensed that the eggs ran a higher risk of perishing, and did not dare to invest many eggs in the risky venture, preferring perhaps to keep some energy in reserve—eggs can be reabsorbed before being laid—for a future attempt, hopefully one with a more secure site.[20]

An interesting example of decision making in a parental species—and one that does not involve predation risk, for a change—is an anecdote reported by Konrad Lorenz, one of the founders of ethology, in his 1952 book *King Solomon's Ring*.[21]

Late one day, Lorenz came to feed a pair of parental jewel cichlids he was keeping in his laboratory. That pair had just about finished retrieving their young for the night—like many cichlids, jewels at dusk gather their free-swimming young a few at a time into their mouth and spit them into a pit so that they can watch over them at night. The female was holding station over the pit full of fry, while the male was dashing back and forth, looking for stragglers. Lorenz dropped a piece of earthworm into the water. The female did not flinch from her guarding post, but the male rushed to the worm, seized it, and started chewing. Then he saw a stray fry swimming by itself away from the pit. Bent on retrieving it, he took it in his already full mouth—and then paused. What to do? To eat or not to eat? To retrieve or not to retrieve? Part of the mouth content had to go to the nest, the other to the stomach. After a few moments, the father found a solution: he spat out both the worm piece and the young. Both sank to the bottom—sinking is a reflex in young cichlid fry being retrieved, and as for the meat, well, that was only gravity. Then the father ate the worm, taking his time and watching the nearby fry. When he was

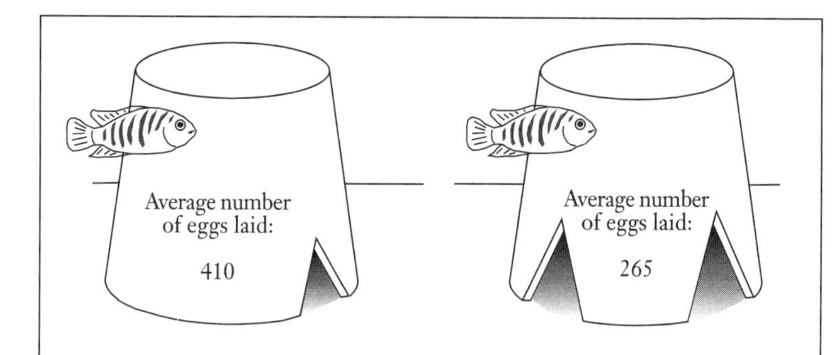

Fig. 11.2. Female convict cichlids may adjust the number of eggs they lay in spawning caves according to the relative safety of the site. Experiments have shown that females lay more eggs inside flowerpots that have only one small opening as opposed to two large openings, the latter being harder to guard against egg predators.

done, he took the fry in his mouth once again and brought it back to its waiting mother.

Nearby students watching the scene spontaneously broke into applause. The decision reached by the fish almost made him look wise.

Animal behavior is an object of fascination, amusement, and wonder. To watch it, one need not face the exacting conditions of fieldwork. From a comfortable chair set up in a dark basement, in front of a well-lit aquarium, one can observe the actions of colorful, resourceful, and not-so-dumb fishes. Smaller than most mammals, less bothered by captivity than birds, esthetically more appealing than insects, fishes represent an ideal subject for behavioral observation. And behavioral experiments can be conducted by anyone with a minimum amount of time and curiosity, a few transparent partitions, crude models, automatic feeders, and maybe a video camera. Many ideas can be tested in a simple manner, with respect for animal welfare. I encourage everyone interested in fishes to try it. At the very least, I trust that the experiments reported here will ensure that readers will never look at a fish—no matter how placid it appears to be—in quite the same way again.

Notes

1. Olfaction

a. Cost of manufacture: B.D. Wisenden and R.J.F. Smith, The effect of physical condition and shoalmate familiarity on proliferation of alarm substance cells in the epidermis of fathead minnows, *Journal of Fish Biology* 50 (1997), 799–808. Spawning season: R.J.F. Smith, Seasonal loss of alarm substance cells in North American cyprinoid fishes and its relation to abrasive spawning behaviour, *Canadian Journal of Zoology* 54 (1976), 1172–82; see also 2230–31 in the same volume and by the same author for the report that although they lose their alarm substance cells, the breeding males retain their fright reaction to the substance released by others. Effect of hunger: G.E. Brown and R.J.F. Smith, Foraging trade-offs in fathead minnows (*Pimephales promelas*, Osteichthyes, Cyprinidae): Acquired predator recognition in the absence of an alarm response, *Ethology* 102 (1996), 776–85; R.J.F. Smith, Effect of food deprivation on the reaction of Iowa darters (*Etheostoma exile*) to skin extract, *Canadian Journal of Zoology* 59 (1981), 558–60. Unresponsive populations: A.E. Magurran, P.W. Irving, and P.A. Henderson, Is there a fish alarm pheromone? A wild study and critique, *Proceedings of the Royal Society of London B* 263 (1996), 1551–56; P.W. Irving and A.E. Magurran, Context-dependent fright reactions in captive European minnows: The importance of naturalness in laboratory experiments, *Animal Behaviour* 53 (1997), 1193–1201; but for a repartee, see R.J.F. Smith, Does one result trump all others? A response to Magurran, Irving and Henderson, *Proceedings of the Royal Society of London B* 264 (1997), 445–50.

b. S.G. Reebs and P.W. Colgan, Nocturnal care of eggs and circadian rhythms of fanning activity in two normally diurnal cichlid fish, *Cichlasoma nigrofasciatum* and *Herotilapia multispinosa*, *Animal Behaviour* 41 (1991), 303–11; R.J. Lavery and S.G. Reebs, Effect of mate removal on current and subsequent parental care in the convict cichlid (Pisces: Cichlidae), *Ethology* 97 (1994), 265–77.

c. J. Caprio, High sensitivity of catfish taste receptors to amino acids, *Comparative Physiology and Biochemistry* 52A (1975), 247–51.

d. For a review on oxygen, see D.L. Kramer, Dissolved oxygen and fish behavior, *Environmental Biology of Fishes* 18 (1987), 81–92. Fanning behavior and oxygen: J.J.A. van Iersel, An analysis of the parental behaviour of the male three-spine stickleback (*Gasterosteus aculeatus* L.), *Behaviour* suppl. 3 (1953), 1–159; P. Sevenster, A causal analysis of a displacement activity (fanning in *Gasterosteus aculeatus* L.), *Behaviour* suppl. 9 (1961), 1–170; S.G. Reebs, F.G. Whoriskey, and G.J. FitzGerald, Diel patterns of fanning activity, egg respiration, and the nocturnal behavior of male three-spined sticklebacks, *Gasterosteus aculeatus* L. (f. *trachurus*), *Canadian Journal of Zoology* 62 (1984), 329–34. Wriggler hanging: S.C. Courtenay and M.H.A. Keenleyside, Wriggler-hanging: A response to hypoxia by brood-rearing *Herotilapia multispinosa* (Teleostei, Cichlidae), *Behaviour* 85 (1983), 183–97.

1. For examples of very small quantities of alarm substance eliciting a response, see B.J. Lawrence and R.J.F. Smith, Behavioral response of solitary fathead minnows, *Pimephales promelas*, to alarm substance, *Journal of Chemical Ecology* 15 (1989), 209–19; R. Poulin, D.J. Marcogliese, and J.D. McLaughlin, Skin-penetrating parasites and the release of alarm substances in juvenile rainbow trout, *Journal of Fish Biology* 55 (1999), 47–53.

2. For reviews of chemical alarm systems in fishes, see W. Pfeiffer, Pheromones in fish and amphibia, in *Pheromones*, ed. M.C. Birch (Amsterdam: North-Holland Publishing Co., 1974), 269–96; W. Pfeiffer, The distribution of fright reaction and alarm substance cells in fishes, *Copeia* (1977), 653–65. R.J.F. Smith, The adaptive significance of the alarm substance—Fright reaction system, in *Chemoreception in Fishes*, ed. T.J. Hara (Amsterdam: Elsevier, 1982), 327–34; R.J.F. Smith, The evolution of chemical alarm signals in fishes, in *Chemical Signals in Vertebrates*, vol. 4, *Ecology, Evolution and Comparative Biology*, ed. D. Duval, D. Müller-Schwarze, and R.M. Silverstein (New York: Plenum Press, 1986), 99–115; R.J.F. Smith, Alarm signals in fishes, *Reviews in Fish Biology and Fisheries* 2 (1992), 33–63.

3. Except in only a few fishes such as piranhas and armored catfishes, alarm systems are thought to be widespread in the fish superorder Ostariophysi, a group that comprises more than 70% of all freshwater species. Recently, however, increasing numbers of non-ostariophysans have also been shown to have similar systems. For examples, see G.E. Brown and R.J.F. Smith, Conspecific skin extracts elicit antipredator responses in juvenile rainbow trout (*Oncorhynchus mykiss*), *Canadian Journal of Zoology* 75 (1997), 1916–22; G.E. Brown and J.-G.J. Godin, Anti-predator responses to conspecific and heterospecific skin extracts

by threespine sticklebacks: Alarm pheromones revisited, *Behaviour* 134 (1997), 1123–34.

4. A. Mathis and R.J.F. Smith, Avoidance of areas marked with a chemical alarm substance by fathead minnows (*Pimephales promelas*) in a natural habitat, *Canadian Journal of Zoology* 70 (1992), 1473–76; D.P. Chivers, B.D. Wisenden, and R.J.F. Smith, The role of experience in the response of fathead minnows (*Pimephales promelas*) to skin extract of Iowa darters (*Etheostoma exile*), *Behaviour* 132 (1995), 665–74.

5. G.E. Brown, D.P. Chivers, and R.J.F. Smith, Localized defecation by pike: A response to labeling by cyprinid alarm pheromone? *Behavioral Ecology and Sociobiology* 36 (1995), 105–10. See also, by the same authors, Fathead minnows avoid conspecific and heterospecific alarm pheromones in the faeces of northern pike, *Journal of Fish Biology* 47 (1995), 387–93 and references therein; and Effects of diet on localized defecation by northern pike, *Esox lucius*, *Journal of Chemical Ecology* 22 (1996), 467–75. Also, from a different research group, R.D. Godard, B.B. Bowers, and C. Wannamaker, Responses of golden shiner minnows to chemical cues from snake predators, *Behaviour* 135 (1998), 1213–28.

6. A. Mathis, D.P. Chivers, and R.J.F. Smith, Chemical alarm signals: Predator-deterrents or predator-attractants? *American Naturalist* 145 (1995), 994–1005; D.P. Chivers, G.E. Brown, and R.J.F. Smith, The evolution of chemical alarm signals: Attracting predators benefits alarm signal senders, *American Naturalist* 148 (1996), 649–59.

7. For a review, see P.W. Sorensen, Hormones, pheromones and chemoreception, in *Fish Chemoreception*, ed. T.J. Hara (London: Chapman and Hall, 1992), 199–228.

8. W.N. Tavolga, Visual, chemical and sound stimuli as cues in the sex discriminatory behavior of the gobiid fish *Bathygobius soporator*, *Zoologica* 41 (1956), 49–64.

9. J.R. Hunter and A.D. Hasler, Spawning association of the redfin shiner, *Notropis umbratilis*, and the green sunfish, *Lepomis cyanellus*, *Copeia* (1965), 265–81.

10. N.R. Liley, Chemical communication in fish, *Canadian Journal of Fisheries and Aquatic Sciences* 39 (1982), 22–35.

11. J.H. Todd, The chemical languages of fishes, *Scientific American* 224 (1971), 98–108.

12. S.G. Reebs and P.W. Colgan, Proximal cues for nocturnal egg care in convict cichlids, *Cichlasoma nigrofasciatum*, *Animal Behaviour* 43 (1992), 209–14.

13. K.A. Jones, Food search behaviour in fish and the use of chemical lures in commercial and sports fishing, in *Fish Chemoreception*, 288–320; also, T. Marui and J. Caprio, Teleost gustation, in *Fish Chemoreception*, 171–98.

14. J.E. Bardach, J.H. Todd, and R. Crickmer, Orientation by taste in fish of the genus *Ictalurus*, *Science* 155 (1967), 1276–78.

15. H. Kleerekoper, The role of olfaction in the orientation of fishes, in *Chemoreception in Fishes*, 201–25.

16. M. Arvedlund and L.E. Nielsen, Do the anemonefish *Amphiprion ocellaris* (Pisces: Pomacentridae) imprint themselves to their host sea anemone *Heteractis magnifica* (Anthozoa: Actinidae)? *Ethology* 102 (1996), 197–211; K. Miyagawa, Experimental analysis of the symbiosis between anemonefish and sea anemones, *Ethology* 80 (1989), 19–46; J.K. Elliott, J.M. Elliott, and R.N. Mariscal, Host selection, location, and association behaviors of anemonefishes in field settlement experiments, *Marine Biology* 122 (1995), 377–89.

17. H. Sweatman, Field evidence that settling coral reef fish larvae detect resident fishes using dissolved chemical cues, *Journal of Experimental Marine Biology and Ecology* 124 (1988), 163–74.

18. M. Halvorsen and O.B. Stabell, Homing behaviour of displaced stream-dwelling brown trout, *Animal Behaviour* 39 (1990), 1089–97.

19. A.D. Hasler and A.T. Scholz, *Olfactory Imprinting and Homing in Salmon* (Berlin: Springer-Verlag, 1983).

20. L. Tosi and C. Sola, Role of geosmin, a typical inland water odour, in guiding glass eel *Anguilla anguilla* (L.) migration, *Ethology* 95 (1993), 177–85.

21. H. Nordeng, A pheromone hypothesis for homeward migration in anadromous salmonids, *Oikos* 28 (1977), 155–59; H. Nordeng, Is the local orientation of anadromous fishes determined by pheromones? *Nature* 233 (1971), 411–13; K.H. Olsén, Kin recognition in fish mediated by chemical cues, in *Fish Chemoreception*, 229–48.

22. M. Takeda and K. Takii, Gustation and nutrition in fishes: Application to aquaculture, in *Fish Chemoreception*, 271–87.

23. D.A. Klaprat, R.E. Evans, and T.J. Hara, Environmental contaminants and chemoreception in fishes, in *Fish Chemoreception*, 321–41.

2. Hearing

1. K. von Frisch, The sense of hearing in fish, *Nature* 141 (1938), 8–11.

2. A.D. Hawkins, The hearing abilities of fish, in *Hearing and Sound Communication in Fishes*, ed. W.N. Tavolga, A.N. Popper, and R.R. Fay (New York: Springer-Verlag, 1981), 109–33; A.D. Hawkins, Underwater sound and fish behaviour, in *Behaviour of Teleost Fishes*, 2nd ed., ed. T.J. Pitcher (London: Chapman and Hall, 1993), 129–69; A. Schuijf and A.D. Hawkins, Acoustic distance discrimination by the cod, *Nature* 302 (1983), 143–44.

3. J.H.S. Blaxter, The swimbladder and hearing, in *Hearing and Sound Communication in Fishes*, 61–70.

4. O. Sand and P.S. Enger, Evidence for an auditory function of the swimbladder in the cod, *Journal of Experimental Biology* 59 (1973), 405–14; R.R. Fay and A.N. Popper, Acoustic stimulation of the ear of the goldfish (*Carassius auratus*), *Journal of Experimental Biology* 61 (1974), 243–60; R.R. Fay and A.N. Popper, Modes of stimulation of the teleost ear, *Journal of Experimental Biology* 62 (1975), 379–87.

5. C.J. Chapman and O. Sand, Field studies of hearing in two species of flatfish, *Pleuronectes platessa* and *Limanda limanda*, *Comparative Biochemistry and Physiology* 47A (1974), 371–85.

6. J.D. Crawford, P. Jacob, and V. Bénech, Sound production and reproductive ecology of strongly acoustic fish in Africa: *Pollimyrus isidori*, Mormyridae, *Behaviour* 134 (1997), 677–725; R.M. Ibara, L.T. Penny, A.W. Ebeling, G. van Dykhuizen, and G. Cailliet, The mating call of the plainfin midshipman fish, *Porichthys notatus*, in *Predators and Prey in Fishes*, ed. D.L.G. Noakes, D.G. Lindquist, G.S. Helfman, and J.A. Ward (The Hague: Dr. W. Junk Publishers, 1983), 205–12; H.E. Winn, J.A. Marshall, and B. Hazlett, Behavior, diel activities, and stimuli that elicit sound production and reactions to sounds in the longspine squirrelfish, *Copeia* (1964), 413–25; J.W. Gerald, Sound production during courtship in six species of sunfish (Centrarchidae), *Evolution* 25 (1971), 75–87; A.A. Myberg, E. Kramer, and P. Heinecke, Sound production by cichlid fishes, *Science* 149 (1965), 555–58; A.A. Myrberg, Ethology of the bicolor damselfish, *Eupomacentrus partitus* (Pisces: Pomacentridae): A comparative analysis of laboratory and field behaviour, *Animal Behaviour Monographs* 5 (1972), 197–283; W.N. Tavolga, The

significance of underwater sounds produced by males of the gobiid fish, *Bathygobius soporator*, *Physiological Zoology* 31 (1958), 259–71; G.-A. Gray and H.E. Winn, Reproductive ecology and sound production of the toadfish, *Opsanus tau*, *Ecology* 42 (1961), 274–82.

7. A.A. Myrberg, Sound communication and interception in fishes, in *Hearing and Sound Communication in Fishes*, 395–425. This is a good review of many of the topics covered in this chapter.

8. A.A. Myrberg, M. Mohler, and J.D. Catala, Sound production by males of a coral reef fish (*Pomacentrus partitus*): Its significance to females, *Animal Behaviour* 34 (1986), 913–23.

9. A.A. Myrberg and J. Spires, Sound discrimination by the bicolor damselfish, *Eupomacentrus partitus*, *Journal of Experimental Biology* 57 (1972), 727–35; E. Spanier, Aspects of species recognition by sound in four species of damselfishes, genus *Eupomacentrus* (Pisces: Pomacentridae), *Zeitschrift für Tierpsychologie* 51 (1979), 301–16.

10. A.A. Myrberg and R.J. Riggio, Acoustically mediated individual recognition by a coral reef fish (*Pomacentrus partitus*), *Animal Behaviour* 33 (1985), 411–16.

11. F. Ladich, Vocalization during agonistic behaviour in *Cottus gobio* L. (Cottidae): An acoustic threat display, *Ethology* 84 (1990), 193–201; W. Valinski and L. Rigley, Function of sound production by the skunk loach *Botia horae* (Pisces, Cobitidae), *Zeitschrift für Tierpsychologie* 55 (1981), 161–72; W.J. Rowland, Sound production and associated behavior in the jewel fish, *Hemichromis bimaculatus*, *Behaviour* 64 (1978), 125–36.

12. J.F. Stout, Sound communication during the reproductive behavior of *Notropis analostanus* (Pisces: Cyprinidae), *American Midland Naturalist* 94 (1975), 296–325.

13. A. Schwarz, The inhibition of aggressive behavior by sound in the cichlid fish, *Cichlasoma centrarchus*, *Zeitschrift für Tierpsychologie* 35 (1974), 508–17; A. Schwarz, Sound production and associated behavior in a cichlid fish, *Cichlasoma centrarchus*. I. Male–male interactions, *Zeitschrift für Tierpsychologie* 35 (1974), 147–56.

14. F. Ladich, Sound characteristics and outcome of contests in male croaking gouramis (Teleostei), *Ethology* 104 (1998), 517–29.

15. D.R. Nelson and R.H. Johnson, Acoustic attraction of Pacific reef sharks: Effect of pulse intermittency and variability, *Comparative Biochemistry*

and Physiology 42A (1972), 85–95; A. Banner, Use of sound in predation by young lemon sharks, *Negaprion brevirostris* (Poey), *Bulletin of Marine Science* 22 (1972), 251–83.

16. Y. Maniwa, Attraction of bony fish, squid and crab by sound, in *Sound Reception in Fish*, ed. A. Schuijf and A.D. Hawkins (Amsterdam: Elsevier, 1976), 271–82.

17. D.J. Colson, S.N. Patek, E.L. Brainerd, and S.M. Lewis, Sound production during feeding in *Hippocampus* seahorses (Syngnathidae), *Environmental Biology of Fishes* 51 (1998), 221–29.

18. *New Scientist*, 13 September 1997, 28–32.

19. D.A. Mann, Z. Lu, and A.N. Popper, A clupeid fish can detect ultrasound, *Nature* 389 (1997), 341. See also J. Astrup and B. Mohl, Discrimination between high and low repetition rates of ultrasonic pulses by the cod, *Journal of Fish Biology* 52 (1998), 205–8.

3. Lateral Line

a. Poem from C. Platt, A.N. Popper, and R.R. Fay, The ear as part of the octavolateralis system, in *The Mechanosensory Lateral Line: Neurobiology and Evolution*, ed. S. Coombs, P. Görner, and H. Münz (New York: Springer-Verlag, 1989), 633–51. Poem reprinted by permission.

1. For a review, see H. Bleckmann, Role of the lateral line in fish behaviour, in *Behaviour of Teleost Fishes*, 2nd ed., ed. T.J. Pitcher (London: Chapman and Hall, 1993), 201–46.

2. T.L. Poulson, Cave adaptation in amblyopsid fishes, *American Midland Naturalist* 70 (1963), 257–90; T. Teyke, Learning and remembering the environment in the blind cave fish *Anoptichthys jordani*, *Journal of Comparative Physiology A* 164 (1989), 655–62.

3. R. Weissert and C. von Campenhausen, Discrimination between stationary objects by the blind cave fish *Anoptichthys jordani* (Characidae), *Journal of Comparative Physiology A* 143 (1981), 375–81; C. von Campenhausen, I. Riess, and R. Weissert, Detection of stationary objects by the blind cave fish *Anoptichthys jordani* (Characidae), *Journal of Comparative Physiology A* 143 (1981),

369–74; E.S. Hassan, On the discrimination of spatial intervals by the blind cave fish (*Anoptichthys jordani*), *Journal of Comparative Physiology A* 159 (1986), 701–10; H. Abdel-Latif, E.S. Hassan, and C. von Campenhausen, Sensory performance of blind Mexican cave fish after destruction of the canal neuromasts, *Naturwissenschaften* 77 (1990), 237–39.

4. S. Dijkgraaf, The functioning and significance of the lateral-line organs, *Biological Reviews* 38 (1962), 51–105.

5. D. Hoekstra and J. Janssen, Non-visual feeding behavior of the mottled sculpin, *Cottus bairdi*, in Lake Michigan, *Environmental Biology of Fishes* 12 (1985), 111–17.

6. J.C. Montgomery and R.C. Milton, Use of the lateral line for feeding in the torrentfish (*Cheimarrichthys fosteri*), *New Zealand Journal of Zoology* 20 (1993), 121–25.

7. J.C. Montgomery, Lateral line detection of planktonic prey, in *The Mechanosensory Lateral Line: Neurobiology and Evolution*, ed. S. Coombs, P. Görner, and H. Münz (New York: Springer-Verlag, 1989), 561–74.

8. J. Janssen, Use of the lateral line and tactile senses in feeding in four Antarctic nototheniid fishes, *Environmental Biology of Fishes* 47 (1996), 51–64.

9. P.S. Enger, A.J. Kalmijn, and O. Sand, Behavioral investigations on the functions of the lateral line and inner ear in predation, in *The Mechanosensory Lateral Line: Neurobiology and Evolution*, 575–87.

10. J.H.S. Blaxter and L.A. Fuiman, Function of the free neuromasts of marine teleost larvae, in *The Mechanosensory Lateral Line: Neurobiology and Evolution*, 481–99.

11. T.J. Pitcher, B.L. Partridge, and C.S. Wardle, A blind fish can school, *Science* 194 (1976), 963–65. Note that in the same way the lateral line can compensate for the lack of vision, vision can compensate for the lack of a lateral line; see B.L. Partridge and T.J. Pitcher, The sensory basis of fish schools: Relative roles of lateral line and vision, *Journal of Comparative Physiology* 135 (1980), 315–25.

12. J.C. Montgomery, C.F. Baker, and A.G. Carton, The lateral line can mediate rheotaxis in fish, *Nature* 389 (1997), 960–63.

13. A.M. Sutterlin and S. Waddy, Possible role of the posterior lateral line in obstacle entrainment by brook trout (*Salvelinus fontinalis*), *Journal of the Fisheries Research Board of Canada* 32 (1975), 2441–46.

14. M. Satou, H.A. Takeuchi, J. Nishii, M. Tanabe, S. Kitamura, N. Okumoto, and M. Iwata, Behavioral and electrophysiological evidences

that the lateral line is involved in the inter-sexual vibrational communication of the hime salmon (landlocked red salmon, *Oncorhynchus nerka*), *Journal of Comparative Physiology A* 174 (1994), 539–49.

15. See H. Bleckmann, Role of the lateral line in fish behaviour, in *Behaviour of Teleost Fishes*, 201–46.

16. Ibid.

4. Electricity and Magnetism

1. In some catfishes, passive electrolocation may also be used to detect naturally occurring voltage gradients within ponds and streams, and they may rely on this information for proper orientation and navigation; see R.C. Peters and F. Bretschneider, Electric phenomena in the habitat of the catfish *Ictalurus nebulosus* LeS., *Journal of Comparative Physiology* 81 (1972), 345–62; R.C. Peters and F. van Wijland, Electro-orientation in the passive electric catfish, *Ictalurus nebulosus* LeS., *Journal of Comparative Physiology* 92 (1974), 273–80.

2. A.J. Kalmijn, The electric sense of sharks and rays, *Journal of Experimental Biology* 55 (1971), 371–83; A.J. Kalmijn, Electric and magnetic field detection in elasmobranch fishes, *Science* 218 (1982), 916–18. See also M. Watt, C.S. Evans, and J.M.P. Joss, Use of electroreception during foraging by the Australian lungfish, *Animal Behaviour* 58 (1999), 1039–45.

3. L.A. Wilkens, D.F. Russell, X. Pei, and C. Gurgens, The paddlefish rostrum functions as an electrosensory antenna in plankton feeding, *Proceedings of the Royal Society of London B* 264 (1997), 1723–29.

4. For a historical account of strongly electric fishes and their impact on human affairs, see C.H. Wu, Electric fish and the discovery of animal electricity, *American Scientist* 72 (1984), 598–607.

5. G. von der Emde, S. Schwarz, L. Gomez, R. Budelli, and K. Grant, Electric fish measure distance in the dark, *Nature* 395 (1998), 890–94; see also 838–39 in the same issue for a comment on that study.

6. H.W. Lissmann and K.E. Machin, Electric receptors in a non-electric fish (*Clarias*), *Nature* 199 (1958), 88–89. See also H.W. Lissmann, Electric location by fishes, *Scientific American* 208 (1963), 50–59.

7. D.L. Meyer, W. Heiligenberg, and T.H. Bullock, The ventral substrate response: A new postural control mechanism in fishes, *Journal of Comparative*

Physiology A 109 (1976), 59–68; A.S. Feng, The role of the electrosensory system in postural control of the weakly electric fish *Eigenmannia virescens*, *Journal of Neurobiology* 8 (1977), 429–37.

8. P. Cain, W. Gerin, and P. Moller, Short-range navigation of the weakly electric fish, *Gnathonemus petersii* L. (Mormyridae, Teleostei), in novel and familiar environments, *Ethology* 96 (1994), 33–45; P. Cain, Navigation in familiar environments by the weakly electric elephantnose fish, *Gnathonemus petersii* L. (Mormyriformes, Teleostei), *Ethology* 99 (1995), 332–49; A.J. Kalmijn, The detection of electric fields from inanimate and animate sources other than electric organs, in *Handbook of Sensory Physiology*, vol. 3, *Electroreceptors and Other Specialized Receptors in Lower Vertebrates*, ed. H. Autrum, R. Jung, W.R. Loewenstein, and D.M. MacKay (New York: Springer-Verlag, 1974), 147–200.

9. For an excellent review, see B. Kramer, *Electrocommunication in Teleost Fishes: Behavior and Experiments* (Berlin: Springer-Verlag, 1990). Note in passing that as is often the case with communication, messages intended for conspecifics can sometimes be intercepted by predators, which can then locate and attack the sender; see S. Hanika and B. Kramer, Electrosensory prey detection in the African sharptooth catfish, *Clarias gariepinus* (Clariidae), of a weakly electric mormyrid fish, the bulldog (*Marcusenius macrolepidotus*), *Behavioral Ecology and Sociobiology* 48 (2000), 218–28.

10. M. Hagedorn and W. Heiligenberg, Court and spark: Electric signals in the courtship and mating of gymnotoid fish, *Animal Behaviour* 33 (1985), 254–65.

11. A. Scheffel and B. Kramer, Electrocommunication and social behaviour in *Marcusenius senegalensis* (Mormyridae, Teleostei), *Ethology* 103 (1997), 404–20; B. Kramer, The attack frequency of *Gnathonemus petersii* towards electrically silent (denervated) and intact conspecifics, and towards another mormyrid (*Brienomyrus niger*), *Behavioral Ecology and Sociobiology* 1 (1976), 425–46.

12. P.K. McGregor and W.M. Westby, Discrimination of individually characteristic electric organ discharges by a weakly electric fish, *Animal Behaviour* 43 (1992), 977–86.

13. T.H. Bullock, R.H. Hamstra, and H. Scheich, The jamming avoidance response of high frequency electric fish, *Journal of Comparative Physiology* 77 (1972), 1–22. See also W. Heiligenberg, *Principles of Electrolocation and Jamming Avoidance in Electric Fish: A Neuroethological Approach* (Berlin: Springer-Verlag, 1977).

14. S. Mann, N.H.C. Sparks, M.M. Walker, and J.L. Kirschvink, Ultrastructure, morphology and organization of biogenic magnetite from

sockeye salmon, *Oncorhynchus nerka:* Implications for magnetoreception, *Journal of Experimental Biology* 140 (1988), 35–49; M.M. Walker, C.E. Diebel, C.V. Haugh, P.M. Pankhurst, J.C. Montgomery, and C.R. Green, Structure and function of the vertebrate magnetic sense, *Nature* 390 (1997), 371–76; C.E. Diebel, R. Proksch, C.R. Green, P. Neilson, and M.M. Walker, Magnetite defines a vertebrate magnetoreceptor, *Nature* 406 (2000), 299–301.

15. For a review, see R. Wiltschko and W. Wiltschko, *Magnetic Orientation in Animals* (Berlin: Springer-Verlag, 1995).

16. G.L. Chew and G.E. Brown, Orientation of rainbow trout (*Salmo gairdneri*) in normal and null magnetic fields, *Canadian Journal of Zoology* 67 (1989), 641–43.

17. A.J. Kalmijn, Experimental evidence of geomagnetic orientation in elasmobranch fishes, in *Animal Migration, Navigation, and Homing*, ed. K. Schmidt-Koenig and W.T. Keeton (Berlin: Springer-Verlag, 1978), 347–53.

18. P.B. Taylor, Experimental evidence for geomagnetic orientation in juvenile salmon, *Oncorhynchus tschawytscha* Walbaum, *Journal of Fish Biology* 28 (1986), 607–23.

19. T.P. Quinn and E.L. Brannon, The use of celestial and magnetic cues by orienting sockeye salmon smolts, *Journal of Comparative Physiology* 147 (1982), 547–52. See also T.P. Quinn, Evidence for celestial and magnetic compass orientation in lake migrating sockeye salmon fry, *Journal of Comparative Physiology* 137 (1980), 243–48; T.P. Quinn and C. Groot, Orientation of chum salmon (*Oncorhynchus keta*) after internal and external magnetic field alteration, *Canadian Journal of Fisheries and Aquatic Sciences* 40 (1983), 1598–1606.

5. Learning

a. G.E. Brown and R.J.F. Smith, Acquired predator recognition in juvenile rainbow trout (*Oncorhynchus mykiss*): Conditioning hatchery-reared fish to recognize chemical cues of a predator, *Canadian Journal of Fisheries and Aquatic Sciences* 55 (1998), 611–17. Also, see B.A. Berejikian, R.J.F. Smith, E.P. Tezak, S.L. Schroder, and C.M. Knudsen, Chemical alarm signals and complex hatchery rearing habitats affect antipredator behavior and survival of chinook salmon (*Oncorhynchus tshawytscha*) juveniles, *Canadian Journal of Aquatic Sciences* 56 (1999), 830–38.

1. H. Gleitman and P. Rozin, Learning and memory, in *Fish Physiology*, vol. 6, *Environmental Relations and Behavior*, ed. W.S. Hoar and D.J. Randall (New York: Academic Press, 1971), 191–278.

2. S.E. Brandon and M.E. Bitterman, Analysis of autoshaping in goldfish, *Animal Learning and Behavior* 7 (1978), 57–62; H.M. Waxman and J.D. McCleave, Auto-shaping in the archer fish (*Toxotes chatareus*), *Behavioral Biology* 22 (1978), 541–44.

3. There is a popular notion that fighting fish view the opportunity to fight as gratifying, but this is a misconception; see T.I. Thompson, Visual reinforcement in Siamese fighting fish, *Science* 141 (1963), 55–57; P.M. Bronstein, Social reinforcement in *Betta splendens:* A reconsideration, *Journal of Comparative and Physiological Psychology* 95 (1981), 943–50.

4. T. Thompson and T. Sturm, Classical conditioning of aggressive display in Siamese fighting fish, *Journal of Experimental Analysis of Behavior* 8 (1965), 397–403.

5. W.N. Tavolga, Visual, chemical and sound stimuli as cues in the sex discriminating behaviour of the gobiid fish, *Bathygobius soporator*, *Zoologica* 41 (1956), 49–64.

6. K.L. Hollis, E.L. Cadieux, and M.M. Colbert, The biological function of Pavlovian conditioning: A mechanism for mating success in the blue gourami (*Trichogaster trichopterus*), *Journal of Comparative Psychology* 103 (1989), 115–21; K.L. Hollis, V.L. Pharr, M.J. Dumas, G.B. Britton, and J. Field, Classical conditioning provides paternity advantage for territorial male blue gouramis (*Trichogaster trichopterus*), *Journal of Comparative Psychology* 111 (1997), 219–25; K.L. Hollis, The biological function of Pavlovian conditioning: The best defense is a good offense, *Journal of Experimental Psychology: Animal Behavior Processes* 10 (1984), 413–25.

7. D.E. Wright and A. Eastcott, Operant conditioning of feeding behaviour and patterns of feeding in thick lipped mullet, *Crenimugil labrosus* (Risso) and common carp, *Cyprinus carpio* (L.), *Journal of Fish Biology* 20 (1982), 625–34; T. Boujard and J.F. Leatherland, Demand-feeding behaviour and diel pattern of feeding activity in *Oncorhynchus mykiss* held under different photoperiod regimes, *Journal of Fish Biology* 40 (1992), 535–44; F.J. Sánchez-Vázquez, J.A. Madrid, and S. Zamora, Circadian rhythms of feeding activity in sea bass, *Dicentrarchus labrax* L.: Dual phasing capacity of diel demand-feeding pattern, *Journal of Biological Rhythms* 10 (1995), 256–66; F.J. Sánchez-

Vázquez, J.A. Madrid, S. Zamora, M. Iigo, and M. Tabata, Demand feeding and locomotor circadian rhythms in the goldfish, *Carassius auratus:* Dual and independent phasing, *Physiology and Behavior* 60 (1996), 665–74.

8. G. Davey, *Ecological Learning Theory* (London: Routledge, 1989).

9. P. Sevenster, Motivation and learning in sticklebacks, in *The Central Nervous System and Fish Behavior*, ed. D. Dingle (Chicago: University of Chicago Press, 1968), 233–45.

10. G.S. Losey and P. Sevenster, Can three-spined sticklebacks learn when to display? Rewarded displays, *Animal Behaviour* 49 (1995), 137–50.

11. D.W. Coble, G.B. Farabee, and R.O. Anderson, Comparative learning ability of selected fishes, *Canadian Journal of Fisheries and Aquatic Sciences* 42 (1985), 791–96.

12. N. Triplett, The educability of the perch, *American Journal of Psychology* 12 (1900–1901), 354–60. For more recent examples, see H.V.S. Peeke, Habituation of a predatory response in the stickleback (*Gasterosteus aculeatus*), *Behaviour* 132 (1995), 1255–66, and references therein.

13. G.S. Helfman and E.T. Schultz, Social transmission of behavioural traditions in a coral reef fish, *Animal Behaviour* 32 (1984), 379–84.

14. M. Anthouard, A study of social transmission in juvenile *Dicentrarchus labrax* (Pisces: Serranidae), in an operant conditioning situation, *Behaviour* 103 (1987), 266–75.

15. For a review, see J.J. Dodson, The nature and role of learning in the orientation and migratory behaviour of fishes, *Environmental Biology of Fishes* 23 (1998), 161–82.

16. K. Warburton, The use of local landmarks by foraging goldfish, *Animal Behaviour* 40 (1990), 500–505; V.A. Braithwaite, J.D. Armstrong, H.M. McAdam, and F.A. Huntingford, Can juvenile Atlantic salmon use multiple cue systems in spatial learning? *Animal Behaviour* 51 (1996), 1409–15. See also R.N. Hughes and C.M. Blight, Two intertidal fish species use visual association learning to track the status of food patches in a radial maze, *Animal Behaviour* 59 (2000), 613–21.

17. D.P. Chivers and R.J.F. Smith, Fathead minnows (*Pimephales promelas*) learn to recognize chemical stimuli from high-risk habitats by the presence of alarm substance, *Behavioral Ecology* 6 (1995), 155–58.

18. L.R. Aronson, Further studies on orientation and jumping behaviour in the Gobiid fish, *Bathygobius soporator*, *Annals of the New York Academy of*

Sciences 188 (1971), 378–92; L.R. Aronson, Orientation and jumping behavior in the gobiid fish *Bathygobius soporator*, *American Museum Novitates* 1486 (1951), 1–22.

19. E.S. Reese, Orientation behavior of butterflyfishes (family Chaetodontidae) on coral reefs: Spatial learning of route specific landmarks and cognitive maps, *Environmental Biology of Fishes* 25 (1989), 79–86.

20. J. Reighard, An experimental field-study of warning coloration in coral reef fishes, Papers from the Tortugas Laboratory, *Carnegie Institution of Washington* 2 (1908), 257–325.

21. K.C. Kruse and B.M. Stone, Largemouth bass (*Micropterus salmoides*) learn to avoid feeding on toad (*Bufo*) tadpoles, *Animal Behaviour* 32 (1984), 1035–39.

22. V. Gotceitas and P. Colgan, Individual variation in learning by foraging juvenile bluegill sunfish (*Lepomis macrochirus*), *Journal of Comparative Psychology* 102 (1988), 294–99; M.I. Croy and R.N. Thompson, The influence of hunger on feeding behaviour and on the acquisition of learned foraging skills by the fifteen-spined stickleback, *Spinachia spinachia* L., *Animal Behaviour* 41 (1991), 161–70; M.I. Croy and R.N. Hughes, The role of learning and memory in the feeding behaviour of the fifteen-spined stickleback, *Spinachia spinachia* L., *Animal Behaviour* 41 (1991), 149–59.

23. P.A. Mackney and R.N. Hughes, Foraging behaviour and memory window in sticklebacks, *Behaviour* 132 (1995), 1241–53.

24. T.J. Ehlinger, Learning and individual variation in bluegill foraging: Habitat-specific techniques, *Animal Behaviour* 38 (1989), 643–58; E.E. Werner, G.G. Mittelbach, and D.J. Hall, The role of foraging profitability and experience in habitat use by the bluegill sunfish, *Ecology* 62 (1981), 116–25.

25. V. Csányi, Ethological analysis of predator avoidance by the paradise fish (*Macropodus opercularis* L.). I. Recognition and learning of predators, *Behaviour* 92 (1985), 227–40. Recognition of predators can also be innate to a certain extent; see R. Gerlai, Can paradise fish (*Macropodus opercularis*, Anabantidae) recognize a natural predator? An ethological analysis, *Ethology* 94 (1993), 127–36.

26. A. Mathis, D.P. Chivers, and R.J.F. Smith, Cultural transmission of predator recognition in fishes: Intraspecific and interspecific learning, *Animal Behaviour* 51 (1996), 185–201.

27. D.P. Chivers and R.J.F. Smith, Chemical recognition of risky habitats is culturally transmitted among fathead minnows, *Pimephales promelas* (Osteichthyes, Cyprinidae), *Ethology* 99 (1995), 286–96.

28. L.M. Dill, The escape response of the zebra danio (*Brachydanyo rerio*). II. The effect of experience, *Animal Behaviour* 22 (1974), 723–30; A.E. Magurran, The inheritance and development of minnow anti-predator behaviour, *Animal Behaviour* 39 (1990), 834–42; B. Olla and M.W. Davis, The role of learning and stress in predator avoidance of hatchery-reared Coho salmon (*Oncorhynchus kisutch*) juveniles, *Aquaculture* 76 (1989), 209–14.

29. W. Goodey and N.R. Liley, The influence of early experience on escape behaviour in the guppy (*Poecilia reticulata*), *Canadian Journal of Zoology* 64 (1986), 885–88; J.J. Tulley and F.A. Huntingford, Paternal care and the development of adaptive variation in anti-predator responses in sticklebacks, *Animal Behaviour* 35 (1987), 1570–72. However, aggressive interactions between stickleback fry do not contribute to the development of antipredator responses later in life; see P.J. Wright and F.A. Huntingford, Agonistic interactions in juvenile sticklebacks (*Gasterosteus aculeatus*) in relation to local predation risk, *Ethology* 94 (1993), 248–56.

30. G.S. Losey, Ecological cues and experience modify interspecific aggression by the damselfish, *Stegastes fasciolatus, Behaviour* 81 (1982), 14–37.

6. Telling Time

a. For reviews, see T.P. Quinn and A.H. Dittman, Fishes, in *Animal Homing*, ed. F. Papi (London: Chapman and Hall, 1992), 145–211; R.J.F. Smith, *The Control of Fish Migration* (Berlin: Springer-Verlag, 1985).

b. S. Thompson, Homing in a territorial reef fish, *Copeia* (1983), 832–34; J.M. Green and R. Fisher, A field study of homing and orientation to the home site in *Ulvaria subbifurcata* (Pisces: Stichaeidae), *Canadian Journal of Zoology* 55 (1977), 1551–56; L.G. Hart and R.C. Summerfelt (1973), cited in T.P. Quinn and A.H. Dittman, Fishes, in *Animal Homing*, 145–211; C.L. Mesing and A.M. Wicker, Home range, spawning migrations, and homing of radio-tagged Florida largemouth bass in two central Florida lakes, *Transactions of the American Fisheries Society* 115 (1986), 286–95; H.R. Carlson and R.E. Haight, Evidence for a home site and homing of adult yellowtail rockfish, *Sebastes flavidus*, *Journal of the Fisheries Research Board of Canada* 29 (1972), 1011–14.

c. A.D. Hasler and W.J. Wisby, The return of displaced largemouth bass and green sunfish to a "home" area, *Ecology* 39 (1958), 289–93; also, R.A. Parker and A.D. Hasler, Movement of some displaced centrarchids, *Copeia* (1959), 11–13.

d. E.S. Titkov, Characteristics of the daily periodicity of wakefulness and rest in the brown bullhead (*Ictalurus nebulosus*), *Journal of Evolutionary Biochemistry and Physiology* 12 (1976), 305–9; I.G. Karmanova, A.I. Belich, and S.G. Lazarev, An electrophysiological study of wakefulness and sleeplike states in fish and amphibians, in *Brain Mechanisms of Behaviour in Lower Vertebrates*, ed. P.R. Laming (Cambridge: Cambridge University Press, 1981), 181–202. See also C.M. Shapiro and H.R. Hepburn, Sleep in a schooling fish, *Tilapia mossambica*, *Physiology and Behavior* 16 (1976), 613–15.

e. C.M. Shapiro, C.J. Clifford, and D. Borsook, Sleep ontogeny in fish, in *Brain Mechanisms of Behaviour in Lower Vertebrates*, 171–80.

f. S.G. Reebs and P.W. Colgan, Nocturnal care of eggs and circadian rhythms of fanning activity in two normally diurnal cichlid fish, *Cichlasoma nigrofasciatum* and *Herotilapia multispinosa*, *Animal Behaviour* 41 (1991), 303–11; see also S.G. Reebs, F.G. Whoriskey, and G.J. FitzGerald, Diel patterns of fanning activity, egg respiration, and the nocturnal behavior of male three-spined sticklebacks, *Gasterosteus aculeatus* L. (f. *trachurus*), *Canadian Journal of Zoology* 62 (1984), 329–34, and references therein.

g. S.G. Reebs, L. Boudreau, P. Hardie, and R. Cunjak, Diel activity patterns of lake chub and other fishes in a stream habitat, *Canadian Journal of Zoology* 73 (1995), 1221–27. For other examples of activity–inactivity cycles breaking down during migration, see B.L. Olla and A.L. Studholme, Comparative aspects of the activity rhythms of tautog, *Tautoga onitis*, Bluefish, *Pomatomus saltatrix*, and Atlantic mackerel, *Scomber scombrus*, as related to their life habits, in *Rhythmic Activity of Fishes*, ed. J.E. Thorpe (London: Academic Press, 1978), 131–51; T.A. Johnston, Downstream movements of young-of-the-year fishes in Catamaran Brook and the Little Southwest Miramichi River, New Brunswick, *Journal of Fish Biology* 51 (1997), 1047–62.

h. N.H.C. Fraser, N.B. Metcalfe, and J.E. Thorpe, Temperature-dependent switch between diurnal and nocturnal foraging in salmon,

Proceedings of the Royal Society of London B 252 (1993), 135–39; N.H.C. Fraser, J. Heggenes, N.B. Metcalfe, and J.E. Thorpe, Low summer temperatures cause juvenile Atlantic salmon to become nocturnal, *Canadian Journal of Zoology* 73 (1995), 446–51. For another example besides salmon, see F.J. Sánchez-Vásquez, M. Azzaydi, F.J. Martinez, S. Zamora, and J.A. Madrid, Annual rhythms of demand-feeding activity in sea bass: Evidence of a seasonal phase inversion of the diel feeding pattern, *Chronobiology International* 15 (1998), 607–22.

1. S.G. Reebs, The anticipation of night by fry-retrieving convict cichlids, *Animal Behaviour* 48 (1994), 89–95.
2. G.S. Helfman, Fish behaviour by day, night, and twilight, in *Behaviour of Teleost Fishes*, 2nd ed., ed. T.J. Pitcher (London: Chapman and Hall, 1993), 479–512.
3. R. Zoufal and M. Taborsky, Fish foraging periodicity correlates with daily changes of diet quality, *Marine Biology* 108 (1991), 193–96.
4. For reviews, see R.E. Spieler, Feeding-entrained circadian rhythms in fishes, in *Rhythms in Fishes*, ed. M.A. Ali (New York: Plenum Press, 1992), 137–47; T. Boujard and J.F. Leatherland, Circadian rhythms and feeding time in fishes, *Environmental Biology of Fishes* 35 (1992), 109–31.
5. R.E. Davis and J.E. Bardach, Time-co-ordinated prefeeding activity in fish, *Animal Behaviour* 13 (1965), 154–62.
6. R.E. Spieler and T.A. Noeske, Effects of photoperiod and feeding schedule on diel variations of locomotor activity, cortisol, and thyroxine in goldfish, *Transactions of the American Fisheries Society* 113 (1984), 528–39. For another example, see S.G. Reebs and M. Lague, Daily food-anticipatory activity in golden shiners: A test of endogenous timing mechanism, *Physiology and Behavior* 70 (2000), 35–43; M. Lague and S.G. Reebs, Phase-shifting the light–dark cycle influences food-anticipatory activity in golden shiners, *Physiology and Behavior* 70 (2000), 55–59; M. Lague and S.G. Reebs, Food-anticipatory activity of groups of golden shiners during both day and night, *Canadian Journal of Zoology* 78 (2000), 886–89.
7. W.J. Zielinski, The influence of daily variation in foraging cost on the activity of small carnivores, *Animal Behaviour* 36 (1988), 239–49.
8. S.G. Reebs, Time–place learning in golden shiners (Pisces: *Notemigonus crysoleucas*), *Behavioural Processes* 36 (1996), 253–62. See also S.G. Reebs,

Time–place learning based on food but not on predation risk in a fish, the inanga (*Galaxias maculatus*), *Ethology* 105 (1999), 361–71.

9. A.D. Hasler, R.M. Horrall, W.J. Wisby, and W. Braemer, Sun-orientation and homing in fishes, *Limnology and Oceanography* 3 (1958), 353–61; H.O. Schwassmann and A.D. Hasler, The role of the sun's altitude in sun orientation of fish, *Physiological Zoology* 37 (1964), 163–78; H.A. Loyacano, Jr., J.A. Chappell, and S.A. Gauthreaux, Sun-compass orientation in juvenile largemouth bass, *Micropterus salmoides, Transactions of the American Fisheries Society* 106 (1977), 77–79; C.P. Goodyear, Terrestrial and aquatic orientation in the starhead topminnow, *Fundulus notti, Science* 168 (1970), 603–5; C. Groot, On the orientation of young sockeye salmon (*Oncorhynchus nerka*) during their seaward migration out of lakes, *Behaviour* suppl. 14 (1965), 1–198; C.P. Goodyear and D.E. Ferguson, Sun-compass orientation in the mosquitofish, *Gambusia affinis, Animal Behaviour* 17 (1969), 636–40; C.P. Goodyear, Learned orientation in the predator avoidance behavior of mosquitofish, *Gambusia affinis, Behaviour* 45 (1973), 191–224; H.E. Winn, M. Salmon, and N. Roberts, Sun-compass orientation by parrot fishes, *Zeitschrift für Tierpsychologie* 21 (1964), 798–812.

10. C.P. Goodyear and D.H. Bennett, Sun compass orientation of immature bluegill, *Transactions of the American Fisheries Society* 108 (1979), 555–59.

11. M. Kavaliers, Seasonal changes in the circadian period of the lake chub, *Couesius plumbeus, Canadian Journal of Zoology* 56 (1978), 2591–96; M. Kavaliers, Social groupings and circadian activity of the killifish, *Fundulus heteroclitus, Biological Bulletin* 158 (1980), 69–76; M. Kavaliers, Circadian activity of the white sucker, *Catostomus commersoni:* Comparison of individual and shoaling fish, *Canadian Journal of Zoology* 58 (1980), 1399–1403; M. Kavaliers, Circadian locomotor activity rhythms of the burbot, *Lota lota:* Seasonal differences in period length and the effect of pinealectomy, *Journal of Comparative Physiology* 36 (1980), 215–18.

12. H. Kabasawa and S. Ooka-Souda, Circadian rhythms in locomotor ac-ty of the hagfish, *Eptatretus burgeri.* IV. The effect of eye ablation, *Zoological ce* 6 (1989), 135–39; S. Ooka-Souda and H. Kabasawa, Circadian rhythms omotor activity in the hagfish, *Eptatretus burgeri.* V. The effect of light on the free-running rhythm, *Zoological Science* 12 (1995), 337–42; bata, M. Minh-Nyo, H. Niwa, and M. Oguri, Circadian rhythm of lo-or activity in a teleost, *Silurus asotus, Zoological Science* 6 (1989), 367–75;

Y. Morita, M. Tabata, K. Ushida, and M. Samejima, Pineal-dependent locomotor activity of lamprey, *Lampetra japonica*, measured in relation to LD cycle and circadian rhythmicity, *Journal of Comparative Physiology A* 171 (1992), 555–62; H.O. Schwassmann, Activity rhythms in gymnotoid electric fishes, in *Rhythmic Activity of Fishes*, ed. J.E. Thorpe (London: Academic Press, 1978), 235–41; H.O. Schwassmann, Biological rhythms, in *Fish Physiology*, vol. 6, ed. W.S. Hoar and D.J. Randall (London: Academic Press, 1971), 371–428.

13. D.R. Nelson and R.H. Johnson, Diel activity rhythms in the nocturnal, bottom-dwelling sharks *Heterodontus francisci* and *Cephaloscyllium ventriosum*, *Copeia* (1970), 732–39.

14. J. Parzefall, Behavioural ecology of cave-dwelling fishes, in *Behaviour of Teleost Fishes*, 573–606.

15. S.J. Northcott, R.N. Gibson, and E. Morgan, On-shore entrainment of circatidal rhythmicity in *Lipophrys pholis* (Teleostei) by natural zeitgeber and the inhibitory effect of caging, *Marine Behaviour and Physiology* 19 (1991), 63–73; E. Morgan and S. Cordiner, Entrainment of a circa-tidal rhythm in the rock-pool blenny *Lipophrys pholis* by simulated wave action, *Animal Behaviour* 47 (1994), 663–69. For a review of older work, see R.N. Gibson, Lunar and tidal rhythms in fish, in *Rhythmic Activity of Fishes*, 201–213.

7. Individual Recognition

a. S.P. Basquill and J.W.A. Grant, An increase in habitat complexity reduces aggression and monopolization of food by zebra fish (*Danio rerio*), *Canadian Journal of Zoology* 76 (1998), 770–72; B.A. de Boer, Factors influencing the distribution of the damselfish *Chromis cyanea* (Poey), Pomacentridae, on a reef at Curacao, Netherlands Antilles, *Bulletin of Marine Science* 28 (1978), 550–65.

b. D.A. Clayton and T.C. Vaughan, Territorial acquisition in the mudskipper *Boleophthalmus boddarti* (Teleostei, Gobiidae) on the mudflats of Kuwait, *Journal of Zoology, London A* 209 (1986), 501–19; D.A. Clayton, Why mudskippers build walls, *Behaviour* 102 (1987), 185–95.

c. A. Ishimatsu, Y. Hishida, T. Takita, T. Kanda, S. Oikawa, T. Takeda, and K.K. Huat, Mudskippers store air in their burrows, *Nature* 391 (1998),

236–37. About the sand mounds of cichlids, see M.I. Taylor, G.F. Turner, R.L. Robinson, and J.R. Stauffer, Jr., Sexual selection, parasites and bower height skew in a bower-building cichlid fish, *Animal Behaviour* 56 (1998), 379–84.

1. H.W. Fricke, Individual partner recognition in fish: Field studies on *Amphiprion bicinctus*, *Naturwissenschaften* 60 (1973), 204–5.

2. S.G. Reebs, Nocturnal mate recognition and nest guarding by female convict cichlids (Pisces, Cichlidae: *Cichlasoma nigrofasciatum*), *Ethology* 96 (1994), 303–12.

3. E. Hert, Individual recognition of helpers by the breeders in the cichlid fish *Lamprologus brichardi* (Poll, 1974), *Zeitschrift für Tierpsychologie* 68 (1985), 313–25.

4. G.C. DeGannes and M.H.A. Keenleyside, Convict cichlid fry prefer the more maternally active of two parental females, *Animal Behaviour* 44 (1992), 525–31; J.E. Cole and J.A. Ward, An analysis of parental recognition by the young of the cichlid fish, *Etroplus maculatus* (Bloch), *Zeitschrift für Tierpsychologie* 27 (1970), 156–76; R.J. Lavery, R.W. Mackereth, R.C. Robilliard, and M.H.A. Keenleyside, Factors determining parental preference of convict cichlid fry, *Cichlasoma nigrofasciatum* (Pisces: Cichlidae), *Animal Behaviour* 39 (1990), 573–81.

5. C. Barnett, The chemosensory responses of young cichlid fish to parents and predators, *Animal Behaviour* 30 (1982), 35–42.

6. T.F. Hay, Filial imprinting in the convict cichlid fish *Cichlasoma nigrofasciatum*, *Behaviour* 65 (1978), 138–60.

7. J.L. Johnsson, Individual recognition affects aggression and dominance relations in rainbow trout, *Oncorhynchus mykiss*, *Ethology* 103 (1997), 267–82; M.R. Morris, L. Gass, and M.J. Ryan, Assessment and individual recognition of opponents in the pigmy swordtails *Xiphophorus nigrensis* and *X. multilineatus*, *Behavioral Ecology and Sociobiology* 37 (1995), 303–10; Á. Miklósi, J. Haller, and V. Csányi, Different duration of memory for conspecific and heterospecific fish in the Paradise fish (*Macropodus opercularis* L.), *Ethology* 90 (1992), 29–36; R.C. Zayan, Défense du territoire et reconnaissance individuelle chez *Xiphophorus* (Pisces, Poecilidae), *Behaviour* 52 (1975), 266–310.

8. J.R. Waas and P.W. Colgan, Male sticklebacks can distinguish between familiar rivals on the basis of visual cues alone, *Animal Behaviour* 47 (1994), 7–13.

9. R.E. Thresher, The role of individual recognition in the territorial behaviour of the threespot damselfish, *Eupomacentrus planifrons*, *Marine Behaviour and Physiology* 6 (1979), 83–93.

10. J.H. Todd, The chemical languages of fishes, *Scientific American* 224 (1971), 98–108.

11. J.H. Todd, J. Atema, and J. Bardach, Chemical communication in social behavior of a fish, the yellow bullhead (*Ictalurus natalis*), *Science* 158 (1967), 672–73. See also M.G. Carr and J.E. Carr, Individual recognition in the juvenile brown bullhead (*Ictalurus nebulosus*), *Copeia* (1985), 1060–62.

12. L.A. Dugatkin and D.S. Wilson, The prerequisites for strategic behaviour in bluegill sunfish, *Lepomis macrochirus*, *Animal Behaviour* 44 (1992), 223–30. These results are somewhat controversial; see J. Lamprecht and H. Hofer, Cooperation among sunfish: Do they have the cognitive abilities? *Animal Behaviour* 47 (1994), 1457–58; C.M. Lombardi and S.H. Hulbert, Sunfish cognition and pseudoreplication, *Animal Behaviour* 52 (1996), 419–22; a reply by Dugatkin and Wilson follows both articles.

13. L.A. Dugatkin and R.C. Sargent, Male–male association patterns and female proximity in the guppy, *Poecilia reticulata*, *Behavioral Ecology and Sociobiology* 35 (1994), 141–45.

14. A.E. Magurran, B.H. Seghers, P.W. Shaw, and G.R. Carvalho, Schooling preferences for familiar fish in the guppy, *Poecilia reticulata*, *Journal of Fish Biology* 45 (1994), 401–6; S.W. Griffiths and A.E. Magurran, Schooling preference for familiar fish vary with group size in a wild guppy population, *Proceedings of the Royal Society of London B* 264 (1997), 547–51; S.W. Griffiths and A.E. Magurran, Familiarity in schooling fish: How long does it take to acquire? *Animal Behaviour* 53 (1997), 945–49; I. Barber and G.D. Ruxton, The importance of stable schooling: Do familiar sticklebacks stick together? *Proceedings of the Royal Society of London B* 267 (2000), 151–55.

15. G.E. Brown and R.J.F. Smith, Fathead minnows use chemical cues to discriminate natural shoalmates from unfamiliar conspecifics, *Journal of Chemical Ecology* 20 (1994), 3051–61; M. de Fraipont and G. Thinès, Responses of the cavefish *Astyanax mexicanus* (*Anoptichthys antrobius*) to the odor of known or unknown conspecifics, *Experientia* 42 (1986), 1053–54.

16. J.A. Brown and P.W. Colgan, Individual and species recognition in centrarchid fishes: Evidence and hypotheses, *Behavioral Ecology and Sociobiology* 19 (1986), 373–79.

17. N.B. Metcalfe and B.C. Thompson, Fish recognize and prefer to shoal with poor competitors, *Proceedings of the Royal Society of London B* 259 (1995), 207–10.

18. D.P. Chivers, G.E. Brown, and R.J.F. Smith, Familiarity and shoal cohesion in fathead minnows (*Pimephales promelas*): Implications for antipredator behaviour, *Canadian Journal of Zoology* 73 (1995), 955–60. But see also S.W. Griffiths, Preferences for familiar fish do not vary with predation risk in the European minnow, *Journal of Fish Biology* 51 (1997), 489–95.

19. T.P. Quinn and C.A. Busack, Chemosensory recognition of siblings in juvenile coho salmon (*Oncorhynchus kisutch*), *Animal Behaviour* 33 (1985), 51–56; T.P. Quinn and T.J. Hara, Sibling recognition and olfactory sensitivity in juvenile coho salmon (*Oncorhynchus kisutch*), *Canadian Journal of Zoology* 64 (1986), 921–25; S. Winberg and K.H. Olsén, The influence of rearing conditions on the sibling odour preference of juvenile Arctic charr, *Salvelinus alpinus* L., *Animal Behaviour* 44 (1992), 157–64. For a review, see G.E. Brown and J.A. Brown, Kin discrimination in salmonids, *Reviews in Fish Biology and Fisheries* 6 (1996), 201–19. Outside of the salmon family, evidence for kin recognition is controversial; see N. Steck, C. Wedekind, and M. Milinski, No sibling odor preference in juvenile three-spined sticklebacks, *Behavioral Ecology* 10 (1999), 493–97, and references therein.

20. G.E. Brown and J.A. Brown, Social dynamics in salmonid fishes: Do kin make better neighbours? *Animal Behaviour* 45 (1993), 863–71; G.E. Brown and J.A. Brown, Do kin always make better neighbours? The effects of territory quality, *Behavioral Ecology and Sociobiology* 33 (1993), 225–31; G.E. Brown and J.A. Brown, Does kin-biased territorial behavior increase kin-biased foraging in juvenile salmonids? *Behavioral Ecology* 7 (1996), 24–29; G.E. Brown, J.A. Brown, and W.R. Wilson, The effects of kinship on the growth of juvenile Arctic charr, *Journal of Fish Biology* 48 (1996), 313–20. See also K.H. Olsén, T. Järvi, and A.-C. Löf, Aggressiveness and kinship in brown trout (*Salmo trutta*) parr, *Behavioral Ecology* 4 (1996), 445–50.

8. Gauging Predators and Adversaries

a. J.C. Abbott, R.L. Dunbrack, and C.D. Orr, The interaction of size and experience in dominance relationships of juvenile steelhead trout (*Salmo gairdneri*), *Behaviour* 92 (1985), 241–53.

b. J.H. Todd, J. Atema, and J.E. Bardach, Chemical communication in social behavior of a fish, the yellow bullhead (*Ictalurus natalis*), *Science* 158 (1967), 672–73.

1. A.E. Magurran and S. Girling, Predator model recognition and response habituation in shoaling minnows, *Animal Behaviour* 34 (1986), 510–18. See also T.J. Pitcher, D.A. Green, and A.E. Magurran, Dicing with death: Predator inspection in minnow shoals, *Journal of Fish Biology* 28 (1986), 439–48.

2. I. Karplus, M. Goren, and D. Algom, A preliminary experimental analysis of predator face recognition by *Chromis caeruleus* (Pisces, Pomacentridae), *Zeitschrift für Tierpsychologie* 58 (1982), 53–65; V. Altbäcker and V. Csányi, The role of eyespots in predator recognition and antipredator behaviour of the paradise fish, *Macropodus opercularis* L., *Ethology* 85 (1990), 51–57.

3. T. Licht, Discriminating between hungry and satiated predators: The response of guppies (*Poecilia reticulata*) from high and low predation sites, *Ethology* 82 (1989), 238–43.

4. G.S. Helfman, Threat-sensitive predator avoidance in damselfish-trumpetfish interactions, *Behavioral Ecology and Sociobiology* 24 (1989), 47–58.

5. K.E. Murphy and T.J. Pitcher, Predator attack motivation influences the inspection behaviour of European minnows, *Journal of Fish Biology* 50 (1997), 407–17.

6. D.F. Fraser and T.N. Mottolese, Discrimination and avoidance reactions towards predatory and nonpredatory fish by blacknose dace, *Rhinichthys atratulus* (Pisces: Cyprinidae), *Zeitschrift für Tierpsychologie* 66 (1984), 89–100.

7. J.-G.J. Godin and S.A. Davis, Who dares, benefits: Predator approach behaviour in the guppy (*Poecilia reticulata*) deters predator pursuit, *Proceedings of the Royal Society of London B* 259 (1995), 193–200. But see also M. Milinski and P. Boltshauser, Boldness and predator deterrence: A critique of Godin and Davis, *Proceedings of the Royal Society of London B* 262 (1995), 103–5; J.-G.J. Godin and S.A. Davis, Boldness and predator deterrence: A reply to Milinski and Boltshauser, *Proceedings of the Royal Society of London B* 262 (1995), 107–12. For another example that predators can gauge which prey are worth pursuing, see J. Krause and J.-G.J. Godin, Influence of prey foraging posture on flight behavior and predation risk: Predators take advantage of unwary prey, *Behavioral Ecology* 7 (1996), 264–71.

8. P.J. Motta, Response by potential prey to coral reef fish predators, *Animal Behaviour* 31 (1983), 1257–59. Some mobbers can actually go as far as nipping the fins of predators, but they wisely do so from behind; see R.G. Hein, Mobbing behavior in juvenile French grunts (*Haemulon flavolineatum*), *Copeia* (1996), 989–91. Usually mobbing is only directed at predators, not at innocuous species; see D. Coates, The discrimination of and reactions towards predatory and non-predatory species of fish by humbug damselfish, *Dascyllus aruanus* (Pisces, Pomacentridae), *Zeitschrift für Tierpsychologie* 52 (1980), 347–54.

9. A.E. Magurran and B.H. Seghers, Population differences in predator recognition and attack cone avoidance in the guppy *Poecilia reticulata*, *Animal Behaviour* 40 (1990), 443–52; W.J. Dominey, Mobbing in colonially nesting fishes, especially the bluegill, *Lepomis macrochirus*, *Copeia* (1983), 1086–88; C.G.M. Paxton, A.E. Magurran, and S. Zschokko, Caudal eyespots on fish predators influence the inspection behaviour of Trinidadian guppies, *Poecilia reticulata*, *Journal of Fish Biology* 44 (1994), 175–77; L.A. Dugatkin and J.-G.J. Godin, Predator inspection, shoaling and foraging under predation hazard in the Trinidadian guppy, *Poecilia reticulata*, *Environmental Biology of Fishes* 34 (1992), 265–76; A.E. Magurran and B.H. Seghers, Predator inspection behaviour covaries with schooling tendency amongst wild guppy, *Poecilia reticulata*, populations in Trinidad, *Behaviour* 128 (1994), 121–34; A.E. Magurran, Predator inspection behaviour in minnow shoals: Differences between populations and individuals, *Behavioral Ecology and Sociobiology* 19 (1986), 267–73; C.B. Brown and K. Warburton, Differences in timidity and escape responses between predator–naive and predator–sympatric rainbowfish populations, *Ethology* 105 (1999), 491–502; D. Külling and M. Milinski, Size-dependent predation risk and partner quality in predator inspection of sticklebacks, *Animal Behaviour* 44 (1992), 949–55; J.D. Reist, Behavioral variation in pelvic phenotypes of brook stickleback, *Culaea inconstans*, in response to predation by northern pike, *Esox lucius*, *Environmental Biology of Fishes* 8 (1983), 255–67; J.-G.J. Godin and K.A. Valdron-Clark, Risk-taking in stickleback fishes faced with different predatory threats, *Ecoscience* 4 (1997), 246–51.

10. S.C. Beeching, Visual assessment of relative body size in a cichlid fish, the oscar, *Astronotus ocellatus*, *Ethology* 90 (1992), 177–86.

11. M. Enquist, T. Ljunberg, and A. Zandor, Visual assessment of fighting ability in the cichlid fish *Nannacara anomala*, *Animal Behaviour* 35 (1987),

1262–64; M. Enquist and S. Jakobsson, Decision making and assessment in the fighting behaviour of *Nannacara anomala* (Cichlidae, Pisces), *Ethology* 72 (1986), 143–53.

12. C.L. Baube, Manipulations of signalling environment affect male competitive success in three-spined sticklebacks, *Animal Behaviour* 53 (1997), 819–33. For similar results in a cichlid, see M.R. Evans and K. Norris, The importance of carotenoids in signaling during aggressive interactions between male firemouth cichlids (*Cichlasoma meeki*), *Behavioral Ecology* 7 (1996), 1–6.

13. J.I. Johnsson and A. Akerman, Watch and learn: Preview of the fighting ability of opponents alters contest behaviour in rainbow trout, *Animal Behaviour* 56 (1998), 771–76.

14. E.R. Keeley and J.W.A. Grant, Visual information, resource value, and sequential assessment in convict cichlid (*Cichlasoma nigrofasciatum*) contests, *Behavioral Ecology* 4 (1993), 345–49.

15. M.H. Figler and D.M. Einhorn, The territorial prior residence effect in convict cichlids (*Cichlasoma nigrofasciatum* Günther): Temporal aspects of establishment and retention, and proximate mechanisms, *Behaviour* 85 (1983), 157–83; G.F. Turner, The fighting tactics of male mouthbrooding cichlids: The effects of size and residency, *Animal Behaviour* 47 (1994), 655–62; M. Itzkowitz, G. Vollmer, and O. Rios-Cardenas, Competition for breeding sites between monogamous pairs of convict cichlids (*Cichlasoma nigrofasciatum*): Asymmetries in size and prior residence, *Behaviour* 135 (1998), 261–67; S. Chellappa, M.E. Yamamoto, M.S.R.F. Cacho, and F.A. Huntingford, Prior residence, body size and the dynamics of territorial disputes between male freshwater angelfish, *Journal of Fish Biology* 55 (1999), 1163–70.

16. T.C.M. Bakker and P. Sevenster, Determinants of dominance in male sticklebacks (*Gasterosteus aculeatus* L.), *Behaviour* 86 (1983), 55–71.

17. D.F. Frey and R.J. Miller, The establishment of dominance relationships in the blue gourami, *Trichogaster trichopterus* (Pallas), *Behaviour* 42 (1972), 8–62; R.C. Francis, Experimental effects on agonistic behavior in the paradise fish, *Macropodus opercularis*, *Behaviour* 85 (1983), 292–313; J.P. Beaugrand, D. Payette, and C. Goulet, Conflict outcome in male green swordtail fish dyads (*Xiphophorus helleri*): Interaction of body size, prior dominance / subordination experience, and prior residency, *Behaviour* 133 (1996), 303–19; J.L. Beacham and J.A. Newman, Social experience and the formation of dominance relationships in the pumpkinseed sunfish, *Lepomis gibbosus*, *Animal Behaviour* 35 (1987),

1560–63. See also J.P. Beaugrand and C. Goulet, Distinguishing kinds of prior dominance and subordination experiences in males of green swordtail fish (*Xiphophorus helleri*), *Behavioural Processes* 50 (2000), 131–42.

18. F.C. Neat, A.C. Taylor, and F.A. Huntingford, Proximate costs of fighting in male cichlid fish: The role of injuries and energy metabolism, *Animal Behaviour* 55 (1998), 875–82. For other examples, see S. Chellappa and F.A. Huntingford, Depletion of energy reserves during reproductive aggression in male three-spined stickleback, *Gasterosteus aculeatus* L., *Journal of Fish Biology* 35 (1989), 315–16; A. Kodric-Brown and P.F. Nicoletto, The relationship between physical condition and social status in pupfish *Cyprinodon pecosensis*, *Animal Behaviour* 46 (1993), 1234–36.

19. F.C. Neat, F.A. Huntingford, and M.M.C. Beveridge, Fighting and assessment in male cichlid fish: The effects of asymmetries in gonadal state and body size, *Animal Behaviour* 55 (1998), 883–91. For another example that combines the effect of size, prior residency, and fighting spirit, see G.W. Barlow, W. Rogers, and N. Fraley, Do Midas cichlids win through prowess or daring? It depends, *Behavioral Ecology and Sociobiology* 19 (1986), 1–8.

9. Mate Choice

a. D.Y. Shapiro, A. Marconato, and T. Yoshikawa, Sperm economy in a coral reef fish, *Thalassoma bifasciatum*, *Ecology* 75 (1994), 1334–44; A. Marconato and D.Y. Shapiro, Sperm allocation, sperm production and fertilization rates in the bucktooth parrotfish, *Animal Behaviour* 52 (1996), 971–80.

b. A. Marconato, V. Tessari, and G. Marin, The mating system of *Xyrichthys novacula:* Sperm economy and fertilization success, *Journal of Fish Biology* 47 (1995), 292–301. See also R.R. Warner, D.Y. Shapiro, A. Marconato, and C.W. Petersen, Sexual conflict: Males with highest mating success convey the lowest fertilization benefits to females, *Proceedings of the Royal Society of London B* 262 (1995), 135–39.

c. For a thorough review, see M. Taborsky, Sneakers, satellites, and helpers: Parasitic and cooperative behavior in fish reproduction, in *Advances in the Study of Behavior*, vol. 23, ed. P.J.B. Slater, J.S. Rosenblatt, C.T. Snowdon, and M. Milinski (San Diego, Calif.: Academic Press, 1994), 1–100.

1. P. Stockley, M.J.G. Gage, G.A. Parker, and A.P. Moller, Sperm competition in fishes: The evolution of testis size and ejaculate characteristics, *American Naturalist* 149 (1977), 933–54. For an exception, see M. Pyron, Testes mass and reproductive mode of minnows, *Behavioral Ecology and Sociobiology* 48 (2000), 132–36.

2. I.M. Côté and W. Hunte, Female redlip blennies prefer older males, *Animal Behaviour* 46 (1993), 203–5; I.M. Côté and W. Hunte, Male and female mate choice in the redlip blenny: Why bigger is better, *Animal Behaviour* 38 (1989), 78–88. For a list of other references on fish preference for larger mates, see A. Kodric-Brown, Mechanisms of sexual selection: Insights from fishes, *Annales Zoologica Fennici* 27 (1990), 87–100.

3. W.J. Rowland, Mate choice and the supernormality effect in female sticklebacks (*Gasterosteus aculeatus*), *Behavioral Ecology and Sociobiology* 24 (1989), 433–38; W.J. Rowland, The ethological basis of mate choice in male threespine sticklebacks, *Gasterosteus aculeatus*, *Animal Behaviour* 38 (1989), 112–20.

4. M.H.A. Keenleyside, R.W. Rangeley, and B.U. Kuppers, Female mate choice and male parental defense behaviour in the cichlid fish *Cichlasoma nigrofasciatum*, *Canadian Journal of Zoology* 63 (1985), 2489–93; D.B. Nuttall and M.H.A. Keenleyside, Mate choice by the male convict cichlid (*Cichlasoma nigrofasciatum;* Pisces, Cichlidae), *Ethology* 95 (1993), 247–56; M.H.A. Keenleyside, Bigamy and mate choice in the biparental cichlid fish *Cichlasoma nigrofasciatum*, *Behavioral Ecology and Sociobiology* 17 (1985), 285–90.

5. M. Perrone, Jr., Mate size and breeding success in a monogamous cichlid fish, *Environmental Biology of Fishes* 3 (1978), 193–201.

6. J.D. Reynolds and M.R. Gross, Female mate preference enhances offspring growth and reproduction in a fish, *Poecilia reticulata*, *Proceedings of the Royal Society of London B* 250 (1992), 57–62.

7. J.F. Downhower and D.B. Lank, Effect of previous experience on mate choice by female mottled sculpins, *Animal Behaviour* 47 (1994), 369–72. See also T.C.M. Bakker and M. Milinski, Sequential female choice and the previous male effect in sticklebacks, *Behavioral Ecology and Sociobiology* 29 (1991), 205–10.

8. C.L. Baube, W.J. Rowland, and J.B. Fowler, The mechanisms of colour-based mate choice in female threespine sticklebacks: Hue, contrast and configurational cues, *Behaviour* 132 (1995), 979–96; I. Barber, S.A. Arnott, V.A. Braithwaite, J. Andrew, W. Mullen, and F.A. Huntingford, Carotenoid-based

sexual coloration and body condition in nesting male sticklebacks, *Journal of Fish Biology* 57 (2000), 777–90; I. Barber, S.A. Arnott, V.A. Braithwaite, J. Andrew, and F.A. Huntingford, Indirect fitness consequences of mate choice in sticklebacks: Offspring of brighter males grow slowly but resist infections, *Proceedings of the Royal Society of London B* 268 (2001), 71–76; U. Candolin, Male–male competition ensures honest signaling of male parental ability in the three-spined stickleback (*Gasterosteus aculeatus*), *Behavioral Ecology and Sociobiology* 49 (2000), 57–61; U. Candolin, The relationship between signal quality and physical condition: Is sexual signalling honest in the three-spined stickleback? *Animal Behaviour* 58 (1999), 1261–67; M. Milinski and T.C.M. Bakker, Female sticklebacks use male coloration in mate choice and hence avoid parasitized males, *Nature* 344 (1990), 330–33. But for exceptions, see V.A. Braithwaite and I. Barber, Limitations to colour-based sexual preferences in three-spined sticklebacks (*Gasterosteus aculeatus*), *Behavioral Ecology and Sociobiology* 47 (2000), 413–16.

9. K.D. Long and A.E. Houde, Orange spots as a visual cue for female mate choice in the guppy (*Poecilia reticulata*), *Ethology* 82 (1989), 316–24.

10. A. Kodric-Brown, Dietary carotenoids and male mating success in the guppy: An environmental component to female choice, *Behavioral Ecology and Sociobiology* 25 (1989), 393–401.

11. P.F. Nicoletto, The relationship between male ornamentation and swimming performance in the guppy, *Poecilia reticulata*, *Behavioral Ecology and Sociobiology* 28 (1991), 365–70; P.F. Nicoletto, Female sexual response to condition-dependent ornaments in the guppy, *Poecilia reticulata*, *Animal Behaviour* 46 (1993), 441–50.

12. A.E. Houde and A.J. Torio, Effect of parasitic infection on male color pattern and female choice in guppies, *Behavioral Ecology* 3 (1992), 346–51.

13. F. Breden and G. Stoner, Male predation risk determines female preference in the Trinidad guppy, *Nature* 329 (1987), 831–33; A.E. Houde, Genetic difference in female choice between two guppy populations, *Animal Behaviour* 36 (1988), 510–16; A.E. Houde and J.A. Endler, Correlated evolution of female mating preferences and male color patterns in the guppy *Poecilia reticulata*, *Science* 248 (1990), 1405–8.

14. A.L. Basolo, Female preference for male sword length in the green swordtail, *Xiphophorus helleri* (Pisces: Poeciliidae), *Animal Behaviour* 40 (1990), 332–38. For an example from another species, see R.J. Bischoff, J.L. Gould,

and D.I. Rubenstein, Tail size and female choice in the guppy (*Poecilia reticulata*), *Behavioral Ecology and Sociobiology* 17 (1985), 253–55.

15. A. Berglund, G. Rosenqvist, and P. Bernet, Ornamentation predicts reproductive success in female pipefish, *Behavioral Ecology and Sociobiology* 40 (1997), 145–50; P. Bernet, G. Rosenqvist, and A. Berglund, Female–female competition affects female ornamentation in the sex-role reversed pipefish *Syngnathus typhle*, *Behaviour* 135 (1998), 535–50, and references therein.

16. M.R. Morris and K. Casey, Female swordtail fish prefer symmetrical sexual signal, *Animal Behaviour* 55 (1998), 33–39; A. Schlüter, J. Parzefall, and I. Schlupp, Female preference for symmetrical vertical bars in male sailfin mollies, *Animal Behaviour* 56 (1998), 147–53; L. Sheridan and A. Pomiankowski, Female choice for spot asymmetry in the Trinidadian guppy, *Animal Behaviour* 54 (1997), 1523–29. But do not miss S.J. Shettleworth, Female mate choice in swordtails and mollies: Symmetry assesment or Weber's law? *Animal Behaviour* 58 (1999), 1139–42.

17. T.C.M. Bakker and W.J. Rowland, Male mating preference in sticklebacks: Effects of repeated testing and own attractiveness, *Behaviour* 132 (1995), 935–49.

18. G.G. Rosenthal, C.S. Evans, and W.L. Miller, Female preference for dynamic traits in the green swordtail, *Xiphophorus helleri*, *Animal Behaviour* 51 (1996), 811–20. For another example of the use of video sequences in the study of mate choice, or more precisely, a demonstration that female sticklebacks prefer a male whose zigzag nuptial dance is executed at a normal or slightly fast speed (the video tape was played back in various degrees of slow or fast motion in addition to the normal speed), see W.J. Rowland, Do female stickleback care about male courtship vigour? Manipulation of display tempo using video playback, *Behaviour* 132 (1995), 951–61.

19. J.D. Reynolds, Should attractive individuals court more? Theory and a test, *American Naturalist* 141 (1993), 914–27. See also A. Kodric-Brown and P.F. Nicoletto, Consensus among females in their choice of males in the guppy *Poecilia reticulata*, *Behavioral Ecology and Sociobiology* 39 (1996), 395–400.

20. G. Stoner and F. Breden, Phenotypic differentiation in female preference related to geographic variation in male predation risk in the Trinidad guppy (*Poecilia reticulata*), *Behavioral Ecology and Sociobiology* 22 (1988), 285–91.

21. P.F. Nicoletto, The influence of water velocity on the display behavior of male guppies, *Poecilia reticulata*, *Behavioral Ecology* 7 (1996), 272–78. For

a field confirmation of these laboratory results, see P.F. Nicoletto and A. Kodric-Brown, The relationship among swimming performance, courtship behavior, and carotenoid pigmentation of guppies in four rivers of Trinidad, *Environmental Biology of Fishes* 55 (1999), 227–35.

22. C.E.J. Kennedy, J.A. Endler, S.L. Poynton, and H. McMinn, Parasite load predicts mate choice in guppies, *Behavioral Ecology and Sociobiology* 21 (1987), 291–95; H. McMinn, Effects of the nematode parasite *Camallanus cotti* on sexual and non-sexual behaviors in the guppy (*Poecilia reticulata*), *American Zoologist* 30 (1990), 245–49; S. Lopez, Acquired resistance affects male sexual display and female choice in guppies, *Proceedings of the Royal Society of London B* 265 (1998), 717–23.

23. R.A. Knapp and J.T. Kovach, Courtship as an honest indicator of male parental quality in the bicolor damselfish, *Stegastes partitus*, *Behavioral Ecology* 2 (1991), 295–300; R.A. Knapp and R.R. Warner, Male parental care and female choice in the bicolor damselfish, *Stegastes partitus:* Bigger is not always better, *Animal Behaviour* 41 (1991), 747–56. See also E. Forsgren, Female sand gobies prefer good fathers over dominant males, *Proceedings of the Royal Society of London B* 264 (1997), 1283–86.

24. R.C. Sargent, Territory quality, male quality, courtship intrusions, and female nest choice in the threespine stickleback, *Gasterosteus aculeatus*, *Animal Behaviour* 30 (1982), 364–74.

25. J.F. Downhower and L. Brown, Mate preferences of female mottled sculpins, *Cottus bairdi*, *Animal Behaviour* 28 (1980), 728–34; A. Kodric-Brown, Determinants of male reproductive success in pupfish (*Cyprinodon pecosensis*), *Animal Behaviour* 31 (1983), 128–37; C. Magnhagen and L. Kvarnemo, Big is better: The importance of size for reproductive success in male *Pomatoschistus minutus* (Pallas) (Pisces, Gobiidae), *Journal of Fish Biology* 35 (1989), 755–63.

26. S. Thompson, Male spawning success and female choice in the mottled triplefin, *Fosterygion varium* (Pisces: Tripterygiidae), *Animal Behaviour* 34 (1986), 580–89. For another example, see C.M. Nelson, Male size, spawning pit size and female mate choice in a lekking cichlid fish, *Animal Behaviour* 50 (1995), 1587–99.

27. R.R. Warner, Female choice of sites versus mates in a coral reef fish, *Thalassoma bifasciatum*, *Animal Behaviour* 35 (1987), 1470–78; R.R. Warner, Male versus female influences on mating-site determination in a coral reef fish, *Animal Behaviour* 39 (1990), 540–48.

28. R.A. Knapp, The influence of egg survivorship on the subsequent nest fidelity of female bicolour damselfish, *Stegastes partitus*, *Animal Behaviour* 46 (1993), 111–21.

29. L.A. Dugatkin, Sexual selection and imitation: Females copy the mate choice of others, *American Naturalist* 139 (1992), 1384–89; L.A. Dugatkin and J.-G.J. Godin, Reversal of female mate choice by copying in the guppy (*Poecilia reticulata*), *Proceedings of the Royal Society of London B* 249 (1992), 179–84.

30. D.L. Lafleur, G.A. Lozano, and M. Sclafani, Female mate-choice copying in guppies, *Poecilia reticulata:* A re-evaluation, *Animal Behaviour* 54 (1997), 579–86; R. Brooks, Copying and the repeatability of mate choice, *Behavioral Ecology and Sociobiology* 39 (1996), 323–29; R. Brooks, Mate choice copying in guppies: Females avoid the place where they saw courtship, *Behaviour* 136 (1999), 411–21; S.E. Briggs, J.-G.J. Godin, and L.A. Dugatkin, Mate-choice copying under predation risk in the Trinidadian guppy (*Poecilia reticulata*), *Behavioral Ecology* 7 (1996), 151–57; L.A. Dugatkin and J.-G.J. Godin, Effects of hunger on mate-choice copying in the guppy, *Ethology* 104 (1998), 194–202; L.A. Dugatkin and J.-G.J. Godin, Female mate copying in the guppy (*Poecilia reticulata*): Age-dependent effects, *Behavioral Ecology* 4 (1993), 289–92; K. Patriquin-Meldrum and J.-G.J. Godin, Do female three-spined sticklebacks copy the mate choice of others? *American Naturalist* 151 (1998), 570–77, and references therein.

31. I. Schlupp, C. Marler, and M.J. Ryan, Benefit to male sailfin mollies of mating with heterospecific females, *Science* 263 (1994), 373–74; I. Schlupp and M.J. Ryan, Male sailfin mollies (*Poecilia latipinna*) copy the mate choice of other males, *Behavioral Ecology* 8 (1997), 104–7. See also J.W.A. Grant and L.D. Green, Mate choice versus preference for actively courting males by female Japanese medaka (*Oryzias latipes*), *Behavioral Ecology* 7 (1996), 165–67.

32. For a list of references, see J.D. Reynolds and J.C. Jones, Female preference for preferred males is reversed under low oxygen conditions in the common goby (*Pomatoschistus microps*), *Behavioral Ecology* 10 (1999), 149–54.

33. T. Goldschmidt, T.C.M. Bakker, and E. Feuth-de Bruijn, Selective copying in mate choice of female sticklebacks, *Animal Behaviour* 45 (1993), 541–47. However, for different results in another population of sticklebacks, see I.G. Jamieson and P.W. Colgan, Eggs in the nests of males and their effect on mate choice in the three-spined stickleback, *Animal Behaviour* 38 (1989), 859–65.

34. A. Marconato and A. Bisazza, Males whose nests contain eggs are preferred by female *Cottus gobio* L. (Pisces, Cottidae), *Animal Behaviour* 34 (1986), 1580–82; S.B.M. Kraak and J.J. Videler, Mate choice in *Aidablennius sphynx* (Teleostei, Blennidae); females prefer nests containing more eggs, *Behaviour* 119 (1991), 243–66.

35. S.B.M. Kraak and G.G. Groothuis, Female preference for nests with eggs is based on the presence of the eggs themselves, *Behaviour* 131 (1994), 189–206.

36. J.-C. Belles-Isles, D. Cloutier, and G.J. FitzGerald, Female cannibalism and male courtship tactics in threespine sticklebacks, *Behavioral Ecology and Sociobiology* 26 (1990), 363–68; Goldschmidt, Bakker, and Feuth-de Bruijn, Selective copying in mate choice of female sticklebacks.

37. J.D. Reynolds and J.C. Jones, Female preference for preferred males is reversed under low oxygen conditions in the common goby (*Pomatoschistus microps*), *Behavioral Ecology* 10 (1999), 149–54.

38. L.M. Unger and R.C. Sargent, Allopaternal care in the fathead minnow, *Pimephales promelas:* Females prefer males with eggs, *Behavioral Ecology and Sociobiology* 23 (1988), 27–32; R.C. Sargent, Allopaternal care in the fathead minnow, *Pimephales promelas:* Stepfathers discriminate against their adopted eggs, *Behavioral Ecology and Sociobiology* 25 (1989), 379–85.

39. S.K. Li and D.H. Owings, Sexual selection in the three-spined stickleback. II. Nest raiding during the courtship phase, *Behaviour* 64 (1978), 298–304. We can also entertain the hypothesis that female attraction to eggs may have directed the evolution of egg-like sexual ornaments in males, such as the fleshy knobs at the tip of dorsal spines in fantail darters or the egg spots on the anal fin of some cichlids; see R.A. Knapp and R.C. Sargent, Egg-mimicry as a mating strategy in the fantail darter, *Etheostoma flabellare:* Females prefer males with eggs, *Behavioral Ecology and Sociobiology* 25 (1989), 321–26; E. Hert, The function of egg-spot in an African mouth-brooding cichlid fish, *Animal Behaviour* 37 (1989), 726–32.

40. P.C. Sikkel, Egg presence and developmental stage influence spawning-site choice by female garibaldi, *Animal Behaviour* 38 (1989), 447–56; P.C. Sikkel, Filial cannibalism in a paternal-caring marine fish: The influence of egg developmental stage and position in the nest, *Animal Behaviour* 47 (1994), 1149–58.

10. Shoal Choice

a. W. Goodey and N.R. Liley, Grouping fails to influence the escape behaviour of the guppy (*Poecilia reticulata*), *Animal Behaviour* 33 (1985), 1032–33.

b. J. Krause, G.D. Ruxton, and D. Rubenstein, Is there always an influence of shoal size on predator hunting success? *Journal of Fish Biology* 52 (1998), 494–501.

c. G. Mittelbach, Group size and feeding rate in bluegills, *Copeia* (1984), 998–1000. See also P. Eklöv, Group foraging versus solitary foraging efficiency in piscivorous predators: The perch, *Perca fluviatilis*, and pike, *Esox lucius*, patterns, *Animal Behaviour* 44 (1992), 313–26.

d. G.W. Barlow, Extraspecific imposition of social groupings among surgeonfishes (Pisces: Acanthuridae), *Journal of Zoology (London)* 174 (1974), 333–40; D.R. Robertson, H.P.A. Sweatman, E.A. Fletcher, and M.G. Cleland, Schooling as a mechanism for circumventing the territoriality of competitors, *Ecology* 57 (1976), 1208–20; S.A. Foster, Group foraging by a coral reef fish: A mechanism for gaining access to defended resources, *Animal Behaviour* 33 (1985), 782–92; S.A. Foster, Acquisition of a defended resource: A benefit of group foraging for the neotropical wrasse, *Thalassoma lucasanum*, *Environmental Biology of Fishes* 19 (1987), 215–22.

e. R.J. Schmitt and S.W. Strand, Cooperative foraging by yellowtail, *Seriola lalandei* (Carangidae), on two species of fish prey, *Copeia* (1982), 714–17.

f. S.G. Reebs, Can a minority of informed leaders determine the foraging movements of a fish shoal? *Animal Behaviour* 59 (2000), 403–9.

g. F.G. Whoriskey, Stickleback distraction displays: Sexual or foraging deception against egg cannibalism? *Animal Behaviour* 41 (1991), 989–95, and references therein.

1. A.E. Magurran, W. Oulton, and T.J. Pitcher, Vigilant behaviour and shoal size in minnows, *Zeitschrift für Tierpsychologie* 67 (1985), 167–78; A.E. Magurran and T.J. Pitcher, Provenance, shoal size and the sociobiology of predator-evasion behaviour in minnow shoals, *Proceedings of the Royal Society of London B* 229 (1987), 439–65. See also J.-G.J. Godin, L.J. Classon, and M.V. Abrahams, Group vigilance and shoal size in a small characin fish, *Behaviour* 104 (1988), 29–40. This effect of shoal size is possible because fish pay attention

to the alarm reaction of their shoalmates; see A.E. Magurran and A. Hingham, Information transfer across fish shoals under predator threat, *Ethology* 78 (1988), 153–58; C.H. Ryer and B.L. Olla, Information transfer and the facilitation and inhibition of feeding in a schooling fish, *Environmental Biology of Fishes* 30 (1991), 317–23; J. Krause, Transmission of fright reaction between different species of fish, *Behaviour* 127 (1993), 37–48.

2. J. Krause and J.-G. J. Godin, Predator preferences for attacking particular prey group sizes: Consequences for predator hunting success and prey predation risk, *Animal Behaviour* 50 (1995), 465–73; S.R.St.J. Neill and J.M. Cullen, Experiments on whether schooling by their prey affects the hunting behaviour of cephalopods and fish predators, *Journal of Zoology (London)* 172 (1974), 549–69.

3. M.C. Hager and G.S. Helfman, Safety in numbers: Shoal size choice by minnows under predatory threat, *Behavioral Ecology and Sociobiology* 29 (1991), 271–76.

4. E.J. Ashley, L.B. Kats, and J.W. Wolfe, Balancing trade-offs between risk and changing shoal size in northern red-belly dace (*Phoxinus eos*), *Copeia* (1993), 540–42.

5. J. Krause, J.-G.J. Godin, and D. Rubenstein, Group choice as a function of group size differences and assessment time in fish: The influence of species vulnerability to predation, *Ethology* 104 (1998), 68–74. For other examples of preference for large shoals, see M.H.A. Keenleyside, Some aspects of the schooling behaviour in fish, *Behaviour* 8 (1955), 183–248; R.W. Tedeger and J. Krause, Density dependence and numerosity in fright stimulated aggregation behaviour of shoaling fish, *Philosophical Transactions of the Royal Society of London B* 350 (1995), 381–90.

6. P.F. Major, Predator–prey interactions in two schooling fishes, *Caranx ignobilis* and *Stolephorus purpureus*, *Animal Behaviour* 26 (1978), 760–77; J.K. Parrish, Comparison of the hunting behavior of four piscine predators attacking schooling prey, *Ethology* 95 (1993), 233–46; M.J. Morgan and J.-G.J. Godin, Antipredator benefits of schooling behaviour in a cyprinodontid fish, the banded killifish (*Fundulus diaphanus*), *Zeitschrift für Tierpsychologie* 70 (1985), 236–46.

7. L. Landeau and J. Terborgh, Oddity and the "confusion effect" in predation, *Animal Behaviour* 34 (1986), 1372–80; C. Theodorakis, Size segrega-

tion and the effect of oddity on predation risk in minnow schools, *Animal Behaviour* 38 (1989), 496–502.

8. E. Ranta, K. Lindstrom, and N. Peuhkuri, Size matters when three-spined sticklebacks go to school, *Animal Behaviour* 43 (1992), 160–62; E. Ranta, S.-K. Juvonen, and N. Peuhkuri, Further evidence for size-assortative schooling in sticklebacks, *Journal of Fish Biology* 41 (1992), 627–30. See also J. Krause, The influence of food competition and predation risk on size-assortative shoaling in juvenile chub (*Leuciscus cephalus*), *Ethology* 96 (1994), 105–16. Another consequence of the oddity effect is that fish that are in a minority within a shoal may have to spend more time looking around for predators and forage less as a consequence; see N. Peuhkiri, Shoal composition, body size and foraging in sticklebacks, *Behavioral Ecology and Sociobiology* 43 (1998), 333–37.

9. N. Peuhkuri, E. Ranta, and P. Seppä, Size-assortative schooling in free-ranging sticklebacks, *Ethology* 103 (1997), 318–24.

10. J.R. Allan and T.J. Pitcher, Species segregation during predator evasion in cyprinid fish shoals, *Freshwater Biology* 16 (1986), 653–59. See also T.J. Pitcher, A.E. Magurran, and J.R. Allan, Size segregative behaviour in minnow shoals, *Journal of Fish Biology* suppl. A, 29 (1986), 83–95.

11. J. Krause and J.-G.J. Godin, Shoal choice in the banded killifish (*Fundulus diaphanus*, Teleostei, Cyprinodontidae): Effects of predation risk, fish size, species composition and size of shoals, *Ethology* 98 (1994), 128–36.

12. J. Krause, J.-G.J. Godin, and D. Brown, Phenotypic variability within and between fish shoals, *Ecology* 77 (1996), 1586–91; J. Krause, J.-G.J. Godin, and D. Brown, Size-assortativeness in multi-species fish shoals, *Journal of Fish Biology* 49 (1996), 221–25.

13. T.J. Pitcher, A.E. Magurran, and I.J. Winfield, Fish in larger shoals find food faster, *Behavioral Ecology and Sociobiology* 10 (1982), 149–51; C.H. Ryer and B.L. Olla, Social mechanisms facilitating exploitation of spatially variable ephemeral food patches in a pelagic marine fish, *Animal Behaviour* 44 (1992), 69–74; T.A. Baird, C.H. Ryer, and B.L. Olla, Social enhancement of foraging on an ephemeral food source in juvenile walleye pollock, *Theragra chalcogramma*, *Environmental Biology of Fishes* 31 (1991), 307–11; N.E. Street and P.J.B. Hart, Group size and patch location by the stoneloach, *Noemacheilus barbatulus*, a non-visually foraging predator, *Journal of Fish Biology* 27 (1985),

785–92; but see also C.H. Ryer and B.L. Olla, Influences of food distribution on fish foraging behaviour, *Animal Behaviour* 49 (1995), 411–18.

14. M.J. Morgan, The effect of hunger, shoal size and the presence of a predator on shoal cohesiveness in bluntnose minnows, *Pimephales notatus* Rafinesque, *Journal of Fish Biology* 32 (1988), 963–71; C.J. Robinson and T.J. Pitcher, The influence of hunger and ration level on shoal density, polarization and swimming speed of herring, *Clupea harengus* L., *Journal of Fish Biology* 34 (1989), 631–33; J.M.L. Richardson, Shoaling in White Cloud Mountain minnows, *Tanichthys albonubes:* Effects of predation risk and prey hunger, *Animal Behaviour* 48 (1994), 727–30; S.M. Sogard and B.L. Olla, The influence of hunger and predation risk on group cohesion in a pelagic fish, walleye pollock *Theragra chalcogramma, Environmental Biology of Fishes* 50 (1997), 405–13.

15. M. Milinski, An evolutionary stable feeding strategy in sticklebacks, *Zeitschrift für Tierpsychologie* 51 (1979), 36–40; W.J. Sutherland, C.R. Townsend, and J.M. Patmore, A test of the ideal free distribution with unequal competitors, *Behavioral Ecology and Sociobiology* 23 (1988), 51–53; M.V. Abrahams, Foraging guppies and the ideal free distribution: The influence of information on patch choice, *Ethology* 82 (1989), 116–26; V. Gotceitas and P. Colgan, Assessment of patch profitability and the ideal free distribution: The significance of sampling, *Behaviour* 119 (1991), 65–76; J.-G.J. Godin and M.H.A. Keenleyside, Foraging on patchily distributed prey by a cichlid fish (Teleostei, Cichlidae): A test of the ideal free distribution theory, *Animal Behaviour* 32 (1984), 120–31; T.C. Grand and J.W.A. Grant, Spatial predictability of resources and the ideal free distribution in convict cichlids, *Cichlasoma nigrofasciatum, Animal Behaviour* 48 (1994), 909–19; T.C. Grand, Foraging site selection by juvenile coho salmon: Ideal free distribution of unequal competitors, *Animal Behaviour* 53 (1997), 185–96.

16. N. van Havre and G.J. FitzGerald, Shoaling and kin recognition in the threespine stickleback (*Gasterosteus aculeatus* L.), *Biology of Behaviour* 13 (1988), 190–201; J. Krause, The influence of hunger on shoal size choice by three-spined sticklebacks, *Gasterosteus aculeatus, Journal of Fish Biology* 43 (1993), 775–80.

17. S.G. Reebs and N. Saulnier, The effect of hunger on shoal choice in golden shiners (Pisces: Cyprinidae, *Notemigonus crysoleucas*), *Ethology* 103 (1997), 642–52.

18. T.J. Pitcher and A.C. House, Foraging rules for group feeders: Forage area copying depends upon food density in shoaling goldfish, *Ethology* 76 (1987), 161–67; J. Krause, Ideal free distribution and the mechanism of patch profitability assessment in three-spined sticklebacks (*Gasterosteus aculeatus*), *Behaviour* 123 (1992), 27–37.

19. S.G. Reebs and B.Y. Gallant, Food-anticipatory activity as a cue for local enhancement in golden shiners (Pisces: Cyprinidae, *Notemigonus crysoleucas*), *Ethology* 103 (1997), 1060–69.

20. L.A. Dugatkin, G.J. FitzGerald, and J. Lavoie, Juvenile three-spined sticklebacks avoid parasitized conspecifics, *Environmental Biology of Fishes* 39 (1994), 215–18; J. Krause and J.-G.J. Godin, Influence of parasitism on shoal choice in the banded killifish (*Fundulus diaphanus*, Teleostei, Cyprinodontidae), *Ethology* 102 (1996), 40–49. Also I. Barber, L.C. Downey, and V.A. Braithwaite, Parasitism, oddity and the mechanism of shoal choice, *Journal of Fish Biology* 53 (1998), 1365–68.

21. J. Krause, The relationship between foraging and shoal position in a mixed shoal of roach (*Rutilus rutilus*) and chub (*Leuciscus leuciscus*): A field study, *Oecologia* 93 (1993), 356–59.

22. J. Krause, D. Bumann, and D. Todt, Relationship between the position preference and nutritional state of individuals in schools of juvenile roach (*Rutilus rutilus*), *Behavioral Ecology and Sociobiology* 30 (1992), 177–80.

23. D. Bumann, J. Krause, and D. Rubenstein, Mortality risk of spatial positions in animal groups: The danger of being in the front, *Behaviour* 134 (1997), 1063–76.

24. J. Krause, The effect of 'Schreckstoff' on the shoaling behaviour of the minnow: A test of Hamilton's selfish herd theory, *Animal Behaviour* 45 (1993), 1019–24.

25. W.A. Tyler, The adaptive significance of colonial nesting in a coral-reef fish, *Animal Behaviour* 49 (1995), 949–66; W.J. Dominey, Anti-predator function of bluegill sunfish nesting colonies, *Nature* 290 (1981), 586–88; M.R. Gross and A.M. MacMillan, Predation and the evolution of colonial nesting in bluegill sunfish (*Lepomis macrochirus*), *Behavioral Ecology and Sociobiology* 8 (1981), 163–74.

26. S.A. Foster, The implications of divergence in spatial nesting patterns in the geminate Caribbean and Pacific sergeant major damselfishes, *Animal Behaviour* 37 (1989), 465–76; Gross and MacMillan, Predation and the evolution of colonial nesting in bluegill sunfish (*Lepomis macrochirus*), 163–74.

11. Making Compromises

1. E.E. Werner, J.F. Gilliam, D.J. Hall, and G.G. Mittelbach, An experimental test of the effects of predation risk on habitat use in fish, *Ecology* 64 (1983), 1540–48; X. He and J.F. Kitchell, Direct and indirect effects of predation on a fish community: A whole-lake experiment, *Transactions of the American Fisheries Society* 119 (1990), 825–35; W.M. Tonn, C.A. Paszkowski, and I.J. Holopainen, Piscivory and recruitment: Mechanisms structuring prey populations in small lakes, *Ecology* 73 (1992), 951–58; L. Jacobsen and S. Berg, Diel variation in habitat use by planktivores in field enclosure experiments: The effect of submerged macrophytes and predation, *Journal of Fish Biology* 53 (1998), 1207–19.

2. C. Brönmark and J.G. Miner, Predator-induced phenotypical change in body morphology in crucian carp, *Science* 258 (1992), 1348–50; C. Brönmark and L.B. Pettersson, Chemical cues from piscivores induce a change in morphology in crucian carp, *Oikos* 70 (1994), 396–402; O.B. Stabell and M.S. Lwin, Predator-induced phenotypic changes in crucian carp are caused by chemical signals from conspecifics, *Environmental Biology of Fishes* 49 (1997), 145–49.

3. V. Gotceitas and P. Colgan, Selection between densities of artificial vegetation by young bluegills avoiding predation, *Transactions of the American Fisheries Society* 116 (1987), 40–49; V. Gotceitas, Variation in plant stem density and its effects on foraging success of juvenile bluegill sunfish, *Environmental Biology of Fishes* 27 (1990), 63–70; V. Gotceitas, Foraging and predator avoidance: A test of a patch choice model with juvenile bluegill sunfish, *Oecologia* 83 (1990), 346–51.

4. C. Magnhagen, Predation risk and foraging in juvenile pink (*Oncorhynchus gorbusha*) and chum salmon (*O. keta*), *Canadian Journal of Fisheries and Aquatic Sciences* 45 (1988), 592–96; C. Magnhagen, Changes in foraging as a response to predation risk in two gobiid fish species, *Pomatoschistus minutus* and *Gobius niger*, *Marine Ecology Progress Series* 49 (1988), 21–26; L.B. Pettersson and C. Brönmark, Trading off safety against food: State dependent habitat choice and foraging in crucian carp, *Oecologia* 95 (1993), 353–57; B. Damsgard and L. Dill, Risk-taking behavior in weight-compensating coho salmon, *Oncorhynchus kisutch*, *Behavioral Ecology* 9 (1998), 26–32.

5. J.F. Gilliam and D. Fraser, Habitat selection under predation hazard: Test of a model with foraging minnows, *Ecology* 68 (1987), 1856–62; S.J. Holbrook and R.J. Schmitt, The combined effects of predation risk and food reward on patch selection, *Ecology* 69 (1988), 125–34; T.J. Pitcher, S.H. Lang, and J.A. Turner, A risk-balancing trade off between foraging rewards and predation hazard in a shoaling fish, *Behavioral Ecology and Sociobiology* 22 (1988), 225–28; M.V. Abrahams and L.M. Dill, A determination of the energetic equivalence of the risk of predation, *Ecology* 70 (1989), 999–1007; M. Kennedy, C.R. Shave, H.G. Spencer, and R.D. Gray, Quantifying the effect of predation risk on foraging bullies: No need to assume an IFD, *Ecology* 75 (1994), 2220–26.

6. R.D. Cerri and D.F. Fraser, Predation and risk in foraging minnows: Balancing conflicting demands, *American Naturalist* 121 (1983), 552–61; D.F. Fraser and R.D. Cerri, Experimental evaluation of predator–prey relationships in a patchy environment: Consequences for habitat use patterns in minnows, *Ecology* 63 (1982), 307–13; R.J. Schmitt and S.J. Holbrook, Patch selection by juvenile black surfperch (Embiotocidae) under variable risk: Interactive influence of food quality and structural complexity, *Journal of Experimental Marine Biology and Ecology* 85 (1985), 269–85; V. Gotceitas and P. Colgan, Behavioural response of juvenile bluegill sunfish to variation in predation risk and food level, *Ethology* 85 (1990), 247–55.

7. J.F. Savino and R.A. Stein, Behavioural interactions between fish predators and their prey: Effects of plant density, *Animal Behaviour* 37 (1989), 311–21; P. Eklöv and L. Persson, The response of prey to the risk of predation: Proximate cues for refuging juvenile fish, *Animal Behaviour* 51 (1996), 105–15.

8. For a review, see M. Milinski, Predation risk and feeding behaviour, in *Behaviour of Teleost Fishes*, 2nd ed., ed. T.J. Pitcher (London: Chapman and Hall, 1993), 285–305.

9. L.M. Dill and A.H.G. Fraser, Risk of predation and the feeding behavior of juvenile coho salmon (*Oncorhynchus kisutch*), *Behavioral Ecology and Sociobiology* 16 (1984), 65–71. See also N.B. Metcalfe, F.A. Huntingford, and J.E. Thorpe, The influence of predation risk on the feeding motivation and foraging strategy of juvenile Atlantic salmon, *Animal Behaviour* 35 (1987), 901–11. In addition to forcing fishes to forage less, predation risk may rob them of the concentration they need to forage efficiently; see N.B. Metcalfe, F.A. Huntingford, and J.E. Thorpe, Predation risk impairs diet selection in juvenile salmon,

Animal Behaviour 35 (1987), 931–33; A.A. Ibrahim and F.A. Huntingford, Laboratory and field studies of the effect of predation risk on foraging in three-spined sticklebacks (*Gasterosteus aculeatus*), *Behaviour* 109 (1989), 46–57; M. Milinski and R. Heller, Influence of a predator on the optimal foraging behaviour of sticklebacks (*Gasterosteus aculeatus* L.), *Nature* 275 (1978), 642–44. Fishes that concentrate too much on getting food may pay dearly for it with their lives; see J.-G.J. Godin and S.A. Smith, A fitness cost of foraging in the guppy, *Nature* 333 (1988), 69–71; M. Milinski, A predator's costs of overcoming the confusion-effect of swarming prey, *Animal Behaviour* 32 (1984), 1157–62.

10. T.C. Grand and L.M. Dill, The effect of group size on the foraging behaviour of juvenile coho salmon: Reduction of predation risk or increased competition? *Animal Behaviour* 58 (1999), 443–51.

11. L.M. Dill and A.H.G. Fraser, Risk of predation and the feeding behavior of juvenile coho salmon (*Oncorhynchus kisutch*), *Behavioral Ecology and Sociobiology* 16 (1984), 65–71. Also V. Gotceitas and J.-G.J. Godin, Foraging under the risk of predation in juvenile Atlantic salmon (*Salmo salar* L.): Effects of social status and hunger, *Behavioral Ecology and Sociobiology* 29 (1991), 255–61; M.V. Abrahams and A. Sutterlin, The foraging and antipredator behaviour of growth-enhanced transgenic Atlantic salmon, *Animal Behaviour* 58 (1999), 933–42.

12. J.A. Endler, Predation, light intensity and courtship behaviour in *Poecilia reticulata* (Pisces: Poeciliidae), *Animal Behaviour* 35 (1987), 1376–85; A.E. Magurran and B.H. Seghers, Risk sensitive courtship in the guppy (*Poecilia reticulata*), *Behaviour* 112 (1990), 194–201.

13. R. Fuller and A. Berglund, Behavioral responses of a sex-role reversed pipefish to a gradient of perceived predation risk, *Behavioral Ecology* 7 (1996), 69–75; E. Forsgren and C. Magnhagen, Conflicting demands in sand gobies: Predators influence reproductive behaviour, *Behaviour* 126 (1993), 125–35; U. Candolin, Predation risk affects courtship and attractiveness of competing threespine stickleback males, *Behavioral Ecology and Sociobiology* 41 (1997), 81–87. See also C. Magnhagen, Sneaking behaviour and nest defence are affected by predation risk in the common goby, *Animal Behaviour* 50 (1995), 1123–28.

14. R. Houtman and L.M. Dill, The influence of substrate color on the alarm response of tidepool sculpins (*Oligocottus maculosus;* Pisces, Cottidae), *Ethology* 96 (1994), 147–54.

15. D.C. Radabaugh, Seasonal colour changes and shifting antipredator tactics in darters, *Journal of Fish Biology* 34 (1989), 679–85.

16. D.L. Kramer, D. Manley, and R. Bourgeois, The effect of respiratory mode and oxygen concentration on the risk of aerial predation in fishes, *Canadian Journal of Zoology* 61 (1983), 653–65.

17. J.H. Gee, Respiratory patterns and antipredator response in the central mudminnow, *Umbra limi*, a continuous, facultative, air-breathing fish, *Canadian Journal of Zoology* 58 (1980), 819–27. For similar considerations in a number of other species, see D.L. Kramer and J.B. Graham, Synchronous air breathing, a social component of respiration in fishes, *Copeia* (1976), 689–97.

18. E.E. DeMartini, Paternal defence, cannibalism and polygamy: Factors influencing the reproductive success of painted greenlings (Pisces, Hexagrammidae), *Animal Behaviour* 35 (1987), 1145–58; C.W. Petersen and K. Marchetti, Filial cannibalism in the Cortez damselfish, *Stegastes rectifraenum*, *Evolution* 43 (1989), 158–68; K. Lindström and R.C. Sargent, Food access, brood size and filial cannibalism in the fantail darter, *Etheostoma flabellare*, *Behavioral Ecology and Sociobiology* 40 (1997), 107–10; R.C. Sargent, Paternal care and egg survival both increase with clutch size in the fathead minnow, *Pimephales promelas*, *Behavioral Ecology and Sociobiology* 23 (1988), 33–37; W. Mrowka, Filial cannibalism and reproductive success in the maternal mouthbrooding cichlid fish *Pseudocrenilabrus multicolor*, *Behavioral Ecology and Sociobiology* 21 (1987), 257–65; R.J. Lavery and M.H.A. Keenleyside, Filial cannibalism in the biparental fish *Cichlasoma nigrofasciatum* (Pisces: Cichlidae) in response to early brood reductions, *Ethology* 86 (1990), 326–38; M.K. Stott and R. Poulin, Parasites and parental care in male upland bullies (Eleotridae), *Journal of Fish Biology* 48 (1996), 283–91.

19. R.J. Lavery and M.H.A. Keenleyside, Parental investment of a biparental cichlid fish, *Cichlasoma nigrofasciatum*, in relation to brood size and past investment, *Animal Behaviour* 40 (1990), 1128–37. For similar examples in other species, see T.S. Carlisle, Parental response to brood size in a cichlid fish, *Animal Behaviour* 33 (1985), 234–38; M.S. Ridgway, The parental response to brood size manipulation in smallmouth bass (*Micropterus dolomieui*), *Ethology* 80 (1989), 47–54; K. Lindström, Effects of costs and benefits of brood care on filial cannibalism in the sand goby, *Behavioral Ecology and Sociobiology* 42 (1998), 101–6; R.M. Coleman, M.R. Gross, and R.C. Sargent, Parental investment decision rules: A test in bluegill sunfish, *Behavioral Ecology and*

Sociobiology 18 (1985), 59–66. And for an interesting question, see A.P. Galvani and R.M. Coleman, Do parental convict cichlids of different sizes value the same brood number equally? *Animal Behaviour* 56 (1998), 541–46. (They do not; small females are more willing to defend small broods than large females are, so everything is relative.)

20. B.D. Wisenden, Female convict cichlids adjust gonadal investment in current reproduction in response to relative risk of brood predation, *Canadian Journal of Zoology* 71 (1993), 252–56.

21. K.Z. Lorenz, *King Solomon's Ring* (New York: Thomas Y. Crowell Co., 1952).

Subject Index

Species Index